INTRODUCTION TO

BASIC ASTRONOMY WITH A PC

J. L. Lawrence

Published by:

Willmann–Bell, Inc.

P.O. Box 35025
Richmond, Virginia 23235
United States of America

Publishers and Booksellers

Serving Astronomers Worldwide
Since 1973

DISCLAIMER

The software which accompanies this book is distributed strictly on an as is basis. The author and publisher specifically disclaim all warranties, expressed or implied, with respect to software and documentation defects, or suitability of the software for a particular purpose.

The diskette is enclosed in a sealed envelope affixed to the inside back cover of this book. Defective diskettes will be replaced free of charge upon receipt of the original diskette within 30 days of purchase. Replacement diskettes are available for $9.95 plus $1.00 handling. See order form at the back of this book for details.

First published 1989

Printed in the United States of America

Library of Congress Cataloging-in-Publication Data.
Lawrence, J. L. (Jackie L.)
Introduction to basic astronomy with a PC / J.L. Lawrence.
p. cm.
Bibliography: p.
Includes index.
ISBN 0-943396-23-9
1. Astronomy–Data processing. 2. Microcomputers. I. Title.
Title: Basic astronomy with a PC.
QB51.3.E4L38 1989 88-5429
520'.28'5536–dc 19 CIP

Table of Contents

Preface

Most introductory astronomy books fall into one of two categories. Books in the first category are essentially descriptive in nature and are almost devoid of any mathematics. Books in the second category approach astronomy from a mathematical standpoint and are filled with formulas and equations which describe celestial motion. This book falls closest to the second category, but it is also an attempt at being a bridge between both categories. While mathematical calculations are central, this book differs from other approaches because it is more oriented toward application of concepts rather than derivation of formulas. Relatively little mathematical skill is assumed throughout. Furthermore, the present book employs the magic of the home computer to guide the reader through the incredible maze of technical details necessary to predict certain astronomical occurrences such as the location of a star or a planet, or predicting the phase of the Moon. By using the computational power of the computer, much of the tediousness is eliminated from the lengthy calculations that are required for virtually every task. When the principles discussed in this book are combined with a desktop microcomputer, the reader has a powerful and stimulating environment for enjoying the wonders of astronomy.

Many books dealing with celestial mathematics assume the reader understands why a calculation is necessary and needs only to be shown the proper formulas. Emphasis here is on understanding what calculations are required, why they are needed, and how all the pieces fit together. The microcomputer is used to demonstrate the topics discussed so that the reader can concentrate on the major concepts rather than getting lost in a myriad of details.

Computer aided instruction is nothing new. This book attempts to go beyond traditional approaches which typically are only programs containing multiple choice questions and are more oriented towards examinations. The approach taken here is to utilize the computer in conjunction with the textbook to serve as an instructor rather than as an examination administrator. Since progress is paced by the reader, the student is free to review a chapter as often as he likes while the computer patiently guides him. Emphasis throughout is placed on learning by doing as well as learning by observing.

All programs referenced in the text are provided on a floppy diskette included with this book. The programs are provided for those wishing to study and modify them although no claim is made that the implementation chosen is the best or most suitable for the problem being solved. In particular, the reader should be forewarned that very little error checking is done during data entry so that the programs, by virtue of brevity, will be clearer and easier to follow. The reader is encouraged to improve and modify the programs. All programs are menu driven and are written in

BASIC for the IBM PC[1] and compatibles.

The topics included in this book are not discussed in lengthy detail. Rather, they are presented as a basis for further study and as a supplement to classroom textbooks. Appendix C contains a list of suggested readings for further study.

Acknowledgements

Although space travel on a limited scale is an accomplished fact, we are still an Earth bound race whose minds yearn to explore the cosmos. Perhaps by the end of the century, some lucky explorers from our planet will have actually visited Mars, Venus, or some other planet within our solar system. Most of us, however, will have to be content with imaginary journeys. This book is for all of us "armchair" explorers who look up at the night sky and are filled with curiosity and want to understand a little more about how the universe we live in appears to operate. As such, it is written not by a professional astronomer nor even by an amateur astronomer, but from the perspective of an amateur "amateur astronomer."

To the staff of Willmann-Bell I owe a special note of thanks. Without their support this project would have been impossible. Writing a book can be a very discouraging undertaking unless the publishing staff personnel are as supportive and professional as those at Willmann-Bell.

I would also like to thank Jean Meeus, Jóse Olivarez, George Lovi, and others who read and commented on the final manuscript. It is both an honor and an humbling experience for individuals of their stature to review the manuscript and offer so many constructive criticisms. Compared to other sciences, fundamental data about celestial objects may change rapidly during the time it takes to publish a book. (For example, the Voyager 2 flyby in January 1986 increased the number of known Uranian moons from 5 to 15. This project began in 1984 with a first draft available in 1985!) While the reviewers and the editorial staff correctly pointed out several errors and suggested updates for much of the data, any mistakes that remain must be attributed to the author.

Most of all, this book is dedicated to you, Mom and Dad, in appreciation for all your years of encouragement and sacrifice. I know the great debt I owe and can only offer gratitude in return. And as always, Mary Ann, your support has been the single most important factor contributing to the completion of this project.

J. L. Lawrence

[1]IBM and IBM PC are registered trademarks of the International Business Machines Corporation.

Chapter 1

Introduction

For as long as man has been staring up into the starry night sky, he has pondered the mysteries of the universe. The majestic beauty of a still, dark night instills a sense of wonder and awe at the vastness of the universe in which we live. Ancient man was often frightened when unexpected events, such as the appearance of a comet, occurred in the sky. Superstitious minds tried to explain the appearance of a comet or the sudden midday darkness during an eclipse by envisioning all sorts of celestial creatures. In later centuries, those who could accurately predict such events were looked upon as heretics and might even be put on trial or executed for practicing black magic.

Tracing the history of astronomy reveals a great deal about the development of man's philosophical and religious thought as well as the development of mathematics and physics. Intellectual giants throughout history made their greatest discoveries and contributions as they attempted to unravel the secrets of the universe. It is only proper, then, to begin our journey by briefly looking at some of the important thinkers and their contributions to the science of astronomy.

Figure 1.1: *This drawing of Stonehenge was made by Inigo Jones in 1621 who wrote the first book devoted solely to Stonehenge. Some authorities now think that this drawing is too regular and that some of its features never existed.*

1.1 A Brief History of Astronomy

It is impossible to say when astronomy really began. Certainly it began before recorded history making it the oldest science. Archaeological evidence suggests that our ancestors placed great emphasis upon celestial events. Examples abound throughout the world. Some believe that Stonehenge near Salisbury, England is the world's oldest astronomical observatory. Stonehenge may have been used to predict eclipses. We can look to the Babylonians and see that they had recorded the position of the planet Jupiter and had derived a calendar based upon astronomical events. Further to the West, the Mayas were intrigued by Venus and left religious monuments that reveal a great deal about their understanding of the planet. Even in modern times, some of our own holidays are based upon celestial events like the phase of the Moon. Easter, for example, is the first Sunday following the full Moon that occurs on or after the Vernal Equinox (around March 21st). In turn, Whitsun and Trinity Sundays are movable dates because they are tied to Easter.

For our purposes, we will begin this brief history by looking at the contributions made by the Greeks beginning in about 600 BC. By the time Greece had become the acknowledged intellectual center of the world, astronomers were already able to predict the positions of the Sun, Moon, and major planets. The Greeks, however, are especially important in world history because of the work they did in so many diverse areas including philosophy, physics, medicine, mathematics, and law. The "father" of many disciplines is from this ancient civilization. The constellations, the zodiac, and even our word "planet" comes from this period.

When the Greeks looked at the night sky, they visualized mythological characters in the stars much as we

Figure 1.2. *This star chart centered on Hercules is from Alexander Jamieson's* A Celestial Atlas *which was published in England in 1822. By the mid-nineteenth century professional atlases were devoid of these mythological figures.*

might visualize animals in cloud formations. Star maps from this period are interesting, and often amusing, due to the artistic license taken by illustrators to help picture characters in the night sky. The Greeks divided the sky into 48 constellations and catalogued the visible stars within each constellation. Stars today are still referred to by the constellation they are in as a quick and easy way to approximately locate their position. Instead of the original 48 constellations, however, modern astronomers have divided the sky into 88 constellations.

As others had done, the Greeks noticed that the stars, while varying with the seasons, appear to be unchanging except for a small number of "stars" that seem to wander through the night sky. They called such wandering stars *planetes.* Thus, our word "planet" comes from this Greek word meaning wanderer. We now know of 9 planets within the solar system while the ancient Greeks only knew of 5 (Mercury, Venus, Mars, Jupiter, and Saturn) but they also classified the Sun and Moon as planets. Notice that the Greeks did not consider Earth to be a planet.

The Greeks also observed that the planets never wander from a very narrow band of the sky that is called the zodiac. The zodiac, only 16° wide, contains the Sun, Moon, and visible planets at all times with the exception of Venus which occasionally wanders outside the zodiac. Astrologers place a great deal of emphasis on the zodiac and the twelve signs of the zodiac. The twelve signs of the zodiac are the constellations Aries, Taurus, Gemini, Cancer, Leo, Virgo, Libra, Scorpius, Sagittarius, Capricornus, Aquarius, and Pisces. While horoscopes can be entertaining, they are largely discredited by present day scientists.

The Greeks have contributed far more to astronomy than simply naming the constellations. The mathematical tools developed by the Greeks allowed them to determine the size of the Earth, distance to the Moon, and distance to the Sun with remarkable accuracy. Such is the power of Euclidean geometry. Euclid, often called the Father of Geometry, started with five basic statements, called postulates, that he assumed as true. The five postulates are:

1. Between any two points, there is exactly one straight line connecting both points.
2. Any straight line can be extended infinitely.
3. It is possible to draw a circle for any given center and radius.
4. All right angles are equal.
5. Given a line m and a point P not on the line, there is exactly one line which passes through point P that is parallel to line m.

These five postulates form the cornerstones of Euclidean geometry. From these simple statements, a complete system of rules can be derived which allow us to do all sorts of useful things including finding the distance between two points.

The Euclidean postulates are intuitively appealing. We accept them as true because of conditioning and because our "common sense" tells us they must be true. However, it is possible to construct a completely consistent geometry which violates the fifth statement! This geometry is called non-Euclidean geometry and plays an important role in modern theories of physics, mathematics, and astronomy. The idea that parallel lines can intersect each other at sufficiently long distances (a consequence of violating postulate 5) is mind-boggling indeed, but even more disturbing is that it can be shown that if the non-Euclidean system is inconsistent, then so is the Euclidean! Fortunately, for "small" distances, Euclidean and non-Euclidean geometry are indistinguishable. "Small" is difficult to define, but distances inside the Milky Way Galaxy are "small" enough. Further discussion of non-Euclidean geometry is beyond the scope of this book. Euclidean geometry will be assumed henceforth.

Applying the principles of Euclidean geometry, Eratosthenes (267–194 BC) calculated the size of the Earth in about 230 BC. He did so by an ingenious experiment. When the Sun is directly overhead, Eratosthenes noted, it does not cast a shadow. But while the Sun casts no

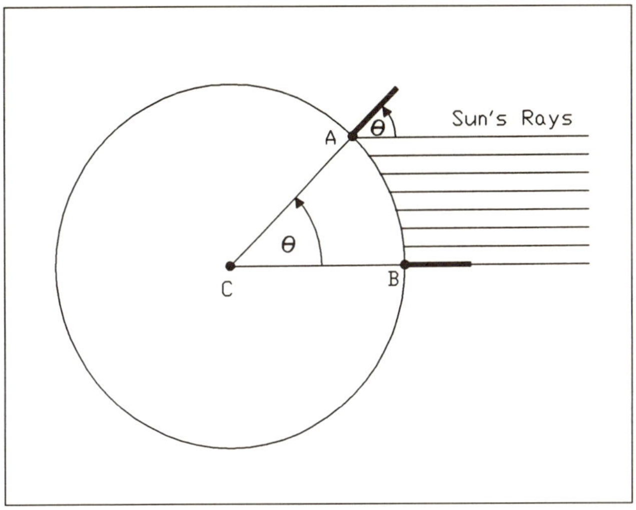

Figure 1.3: *The Earth's circumference can be determined if you know the distance between two points (A and B) that are on a North–South line and the angle of the Sun's shadow at point A when, at the same instant, it casts no shadow at point B.*

shadow in one location, at that exact same instant it will cast a shadow at a location some distance away to the North or to the South. The experiment is illustrated in Figure 1.3. Sticks were driven into the ground at points A and B such that they are perpendicular to the ground. Point A represents the town of Alexandria, B represents the town of Syene, and C represent's the Earth's center. When the Sun is directly overhead at Syene, the Syene stick will cast no shadow while the one at Alexandria will. The angle θ is the angle made by the Sun's rays and the stick at Alexandria which, by the rules of geometry, is the same as $\angle ACB$. Now if θ can be determined as well as the distance from Syene to Alexandria, it is a simple matter to determine the circumference and radius of the Earth.

Before presenting the results of the experiment, note three important details. First, Eratosthenes had to assume the Earth is round. This was nearly 2000 years before Columbus! For some reason, this idea lay forgotten for many years and had to be "rediscovered." Secondly, for the experiment to yield accurate results, Alexandria would have to lie on the same meridian as Syene. For this to be true, Alexandria would have to be directly North of Syene. This is not quite the case but the resulting error is fairly small. Finally, Eratosthenes would need some way of being in Alexandria and knowing that the Sun was directly overhead in Syene. During the summer solstice, the Sun would be almost exactly overhead at noon. Thus, on the date of the summer solstice (around the middle of June), θ could be measured in Alexandria at noon. As it turns out, the Sun was not exactly overhead for Syene, but the error incurred was largely canceled out by the fact that Alexandria and Syene do not lie on the same meridian.

Eratosthenes measured θ to be 7.2° while the actual value is about 7.08°. Assuming that the Earth is perfectly round, the distance Eratosthenes measured from Syene to Alexandria represents a fractional portion of the total circumference of a circle whose radius is the same as that of the Earth. Since there are 360° in a circle, this fraction is calculated as

$$\frac{7.2^\circ}{360^\circ} = \frac{1}{50}. \tag{1.1.1}$$

Therefore, to determine the circumference of the Earth, Eratosthenes merely needed to multiply the distance from Syene to Alexandria by 50. A circle's circumference is given by $2\pi r$, where r is the radius of the circle and π is approximately 3.14159, so the diameter of the Earth is given by dividing its circumference by π. With this approach, Eratosthenes estimated the diameter of the Earth as 7850 miles compared to a modern estimate of about 7900 miles. An error of less than 1% is remarkable considering the primitive tools with which Eratosthenes had to work.

Aristarchus, another Greek, performed the remarkable feat of obtaining an estimate of the distance from the Earth to the Moon and an estimate of the distance from the Earth to the Sun. The lunar distance can be approximated during a lunar eclipse by measuring the apparent diameter of the Earth's shadow on the Moon. Using this figure and the apparent diameter of the Moon, Aristarchus arrived at a ratio representing the Moon's size relative to the Earth. Then, given the diameter of the Earth, Aristarchus could calculate how far away the Moon had to be for the Earth's shadow to be of the size measured during the eclipse.

Going further, Figure 1.4 illustrates how Aristarchus went about estimating the distance to the Sun. When the Moon appears as a half Moon, the angle formed by the Sun, Moon, and Earth must be 90°. This is $\angle SME$ in Figure 1.4. $\angle MES$ can be determined by observation while $\angle MSE$ can be deduced given the other two angles in the triangle. From this data, Aristarchus could compute the distance to the Sun relative to the distance to the Moon. He measured $\angle MES$ to be about 87° and concluded that the Sun must be 20 times farther away than the Moon. Then, using an estimate of the Earth's diameter (made possible by Eratosthenes), Aristarchus arrived at an Earth-Sun distance of 4–5 million miles. Unfortunately, the measurement of $\angle MES$ was not precise enough to get a good estimate of the distance to the Sun. But Aristarchus had managed to make possible the first determination of the scale of our solar system.

Aristarchus believed that the Earth moves around the Sun. Based upon his calculation of the scale of

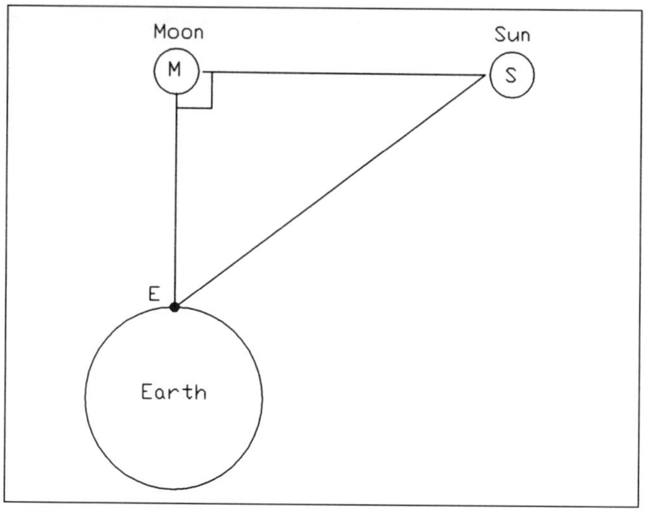

Figure 1.4: *Once the diameter of the Earth was known, Aristarchus was able to estimate the relative distance between the Sun and Earth by calculating the angles between the Sun, Moon and Earth.*

the solar system, he came to a bold conclusion. If the Earth does revolve around the Sun, there can be only one explanation for the apparent unchanging background of stars. The stars must be very far away. But, primarily due to the influence of Aristotle and Ptolemy, the idea of very distant stars and a Sun-centered universe fell into disfavor.

Aristotle (384–322 BC) is perhaps the greatest of the ancient Greek philosophers. In his book *Organon*, Aristotle wrote in favor of an Earth-centered universe. This view was to prevail for many centuries even though others, notably Aristarchus, disagreed. Besides favoring a geocentric (Earth-centered) system, Aristotle described the motion of the planets in terms of celestial spheres. That is, he pictured the planets, Sun, Moon, and stars as being attached to spheres which had the Earth at their center. Thus, the universe was likened to an immense hollow ball with smaller balls inside in a layered fashion. Celestial bodies were thought of as moving in gigantic circles. While many of Aristotle's ideas were discarded much earlier, belief in the existence of the ether did not disappear until the 1900's.

Aristotle was incorrect in many of his theories. As an example, he taught that falling bodies will fall at a speed proportional to their weight. Thus, he explained, a rock falls faster than a feather and a large rock falls faster than a small rock. Aristotle was so highly respected that this incorrect notion was held as unquestionable truth for more than 1500 years. Galileo disproved the theory in the famous experiment in which he dropped two balls of unequal weight from the Leaning Tower of Pisa. When both balls reached the ground at the same time, it is said that some of the witnesses there still refused to believe what they had just seen. Such was the respect for

Aristotle. Still, incorrect theories are often better than no theory at all because they start man thinking and prompt experimentation.

From the point of view of physics and astronomy, Aristotle's most important contribution was his theory of motion. While incorrect, this was the first organized attempt to predict how and why an object would move. In Aristotle's view, motion is one of two types: natural motion or violent motion. Natural motion, Aristotle reasoned, occurs because of the "nature" of the object. In the case of so-called natural motion, objects will move whenever they are out of their natural place. For example, a rock will fall to the ground unless supported because a rock is of the Earth and resting on the ground is its natural state. Smoke, which is mostly air, will rise because that is its natural state. Larger objects will try harder to return to their natural state and hence move faster. With the exception of celestial objects, Aristotle conjectured that the natural state for all objects is to be at a state of rest.

Aristotle thought of natural motion as being either straight up and down or circular. He reasoned that all objects on the Earth will move straight up and down while objects in the heavens will move in circles. From a philosophical viewpoint, this was as it should be because the heavens were considered perfect as were circles. Nothing could be more natural than for the celestial bodies to be embedded in spheres which move in "divine" circles.

All other motion in Aristotle's view fell into the category of violent motion. Violent motion supposedly results from an external force either pushing or pulling an object. As soon as the external force is removed, an object will stop moving. This certainly seemed to fit the facts because anyone could see that a cart will stop moving if the ox refuses to pull it!

To summarize, Aristotle favored a geocentric universe composed of celestial spheres in which all objects move in proportion to their weight and desire to be in their natural state. Furthermore, excepting celestial objects, an object will remain at rest until acted upon by an external force and will stop as soon as the force is removed. Several centuries would pass before Newton would change how mankind thought about motion and explain what Aristotle did not: *why* do the celestial bodies move?

The last major contribution in astronomy by the Greek civilization was the book *The Almagest* written by Ptolemy (100–178 AD). This book contains a catalogue of the visible stars within each constellation, their location, and their brightness. Also included are observations of the Sun, Moon, and planets. Most importantly, Ptolemy developed the concept of epicycles which were

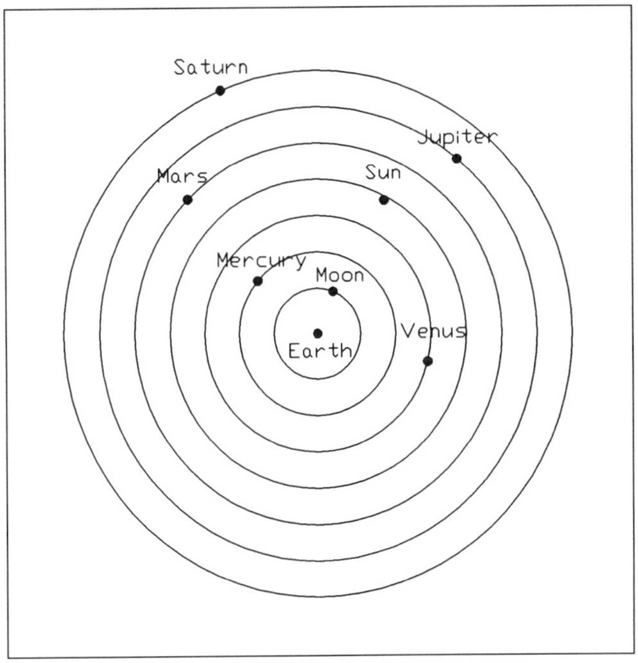

Figure 1.5: *Aristotle, among others, thought the Earth was the center of the universe and that all objects in the heavens moved in circles around the Earth.*

to be an accepted idea in planetary motion for almost 1500 years.

Figure 1.5 is a representation of Aristotle's view of the solar system. The center dot represents the Earth while each of the circles represents a cross-section of a sphere. In order, the dots on the circles represent the Moon, Mercury, Venus, the Sun, Mars, Jupiter, and Saturn. When compared to actual observations, Aristotle's view left the ancients with a baffling mystery. Some of the planets appeared to change direction and for a time move backwards. This backward motion is called retrograde motion and could not be explained using Aristotle's geocentric model.

Ptolemy tried to solve the problem with the model shown in Figure 1.7. The Earth is again assumed to be the center as in Aristotle's view. For clarity, only the Earth and Venus are shown in Figure 1.7. Note that Venus is pictured as revolving in a circle around a point that is also revolving around the Earth. The dashed circle shown in Figure 1.7 is called the deferent while the orbit for Venus is called an epicycle. So, if the center of the orbit of Venus moves in a circle around the Earth while Venus itself revolves in a circle around this moving point, Venus will indeed retrograde periodically as confirmed by observation. Epicycles were added for the other planets exhibiting retrograde motion to form a geocentric model of the solar system. While Ptolemy's addition of epicycles improved upon the model proposed

Figure 1.6: *Ptolemy developed the concept of epicycles to describe the motion of the planets. This idea was to survive for more than 1500 years.*

by Aristotle, it was still inaccurate in many respects.

Due to political and religious upheavals throughout the world, few advances were to be made in astronomy (or any of the sciences) until Copernicus (1473–1543 AD). With the fall of the Greek and Roman empires, much of the learning of the ancient world was lost as mankind slipped into the Dark Ages (500–900 AD). Belief in a geocentric universe became an unquestionable "truth" as well as belief that the Earth was flat. Dissenting views not only ran counter to popular opinion, but ran against established religious doctrine. The first major figure to dissent was Copernicus.

Copernicus, known as the Father of Modern Astronomy, is best remembered for his book *De Revolutionibus Orbium Coelestium* in which he put forth his theories on planetary motion. Although he retained the idea of epicycles, Copernicus revived the notion of a moving Earth. As already pointed out, the concept of a heliocentric (Sun-centered) universe is at least as old as the Greek civilization. Initial reception of the idea was no better for Copernicus than it had been for the Greeks. In fact, Copernicus hesitated publishing his theories for fear of church officials and might not have done so at all had it not been for prodding from his friends. The first copy of the work reached Copernicus on the day he died.

The world was in religious turmoil during the lifetime of Copernicus. It was during this period of world history that Martin Luther began the Reformation Movement.

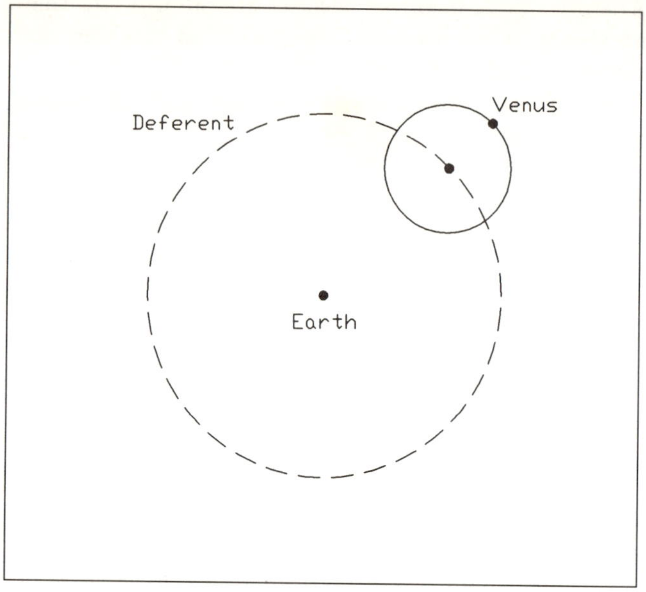

Figure 1.7: *When viewed from Earth, planets can appear to move "backwards" during part of their orbit. Using Aristotle's model of the universe, this was impossible. This retrograde movement was rationalized by Ptolemy as epicycles.*

Figure 1.8: *So controversial was Copernicus' major contribution to astronomy* De Revolutionibus Orbium Coelestium *that he only allowed it to be published when he was near death.*

But Protestant and Catholic alike were united in their opposition to the Copernican view of the solar system. Copernicus was probably justified in his fear of public opinion for his book was banned by the Catholics for almost two hundred years. Others who disagreed with popular opinion would not be so fortunate.

During the Renaissance Period, many of the Aristotelian ideas were questioned and denied by brave individuals. But to disagree with Aristotle was to contradict established religious doctrine. The price was often high as in the case of Giordano Bruno. Bruno believed that the stars are really celestial bodies like our Sun except much farther away. He went so far as to suggest that because of the great stellar distances, the universe might in fact be infinite. For refusing to recant these heretical views, Bruno was burned at the stake in 1600 AD. Bruno's martyrdom undoubtedly influenced his contemporary, Galileo.

Galileo (1564–1642) made several important contributions to early astronomy. He was the first astronomer to use the newly invented telescope to observe the heavens. Among other things, Galileo observed lunar mountains and the rings of Saturn, discovered sunspots, and discovered that the planet Jupiter has moons. His observations led him to the same conclusion as Copernicus; the Earth revolves around the Sun.

Perhaps Galileo's most significant gift to science was his approach to scientific experiments. For this Galileo is given credit for being the Father of the Scientific Method. The essence of the scientific method can be summarized in three steps:

1. Formulate a statement of the problem.

2. Formulate a hypothesis which attempts to explain the problem and predicts the results of experiments.

3. Perform experiments to verify the hypothesis.

Scientists before Galileo often performed experiments, but Galileo was the first to systematically employ the scientific method throughout his work.

Galileo seemed to particularly enjoy using the scientific method to discredit commonly held ideas. His experiment at the Tower of Pisa, as described above, disproved Aristotle's assertion that bodies fall at a rate proportional to their weight. By observing the motion of balls on inclined planes, Galileo attacked the Aristotelian views on motion. Specifically, Galileo was able to show that once a body is in motion, it will continue moving in a straight line unless acted upon by an external force. Thus, Aristotle's concept of how the celestial bodies move was in serious doubt. The man, Newton, who would successfully explain why the planets move was born the year that Galileo died.

When Galileo was about fifty years old, he was warned that he must not teach nor advocate that the Earth is in motion around the Sun. But, ten years later,

Figure 1.9: *Galileo was probably the first person to point a telescope skyward and make meaningful observations which led him to agree with Copernican theory.*

Galileo published *Dialogue on the Chief Systems of the World* in which he affirmed his support of the Copernican theory. For this heresy, Galileo was soon arrested and given the choice of recanting or facing martyrdom. Galileo chose to recant, perhaps recalling the earlier fate of Bruno. Even though he recanted, Galileo was to spend the rest of his life under house arrest.

Galileo may have been the first astronomer to use a telescope for observations, but one of his contemporaries performed more exacting observations *without* a telescope. Tycho Brahe (1546–1601) built an observatory to make very precise measurements of the Sun, Moon, planets, and stars. For about twenty years Tycho and his assistants collected the data. He concluded from the data that the planets revolve around the Sun but that the Sun revolves around a stationary Earth. Even though the reality is that the Earth moves about the Sun, it is sometimes more convenient to make use of this mathematical equivalence and assume a stationary Earth.

One of Tycho's assistants was Johannes Kepler (1571–1630). Upon Tycho's death, Kepler inherited the wealth of astronomical data collected in Tycho's observatory. Without the data collected by Tycho and his

Figure 1.10: *Using data gathered by Tycho Brahe, Johannes Kepler developed laws of planetary movement.*

assistants, Kepler would probably have never been able to deduce his three laws of planetary motion.

Ever since at least the ancient Greeks, it was assumed that celestial motion was in the form of circles. However, Kepler noted that circular orbits did not match the observed data with sufficient accuracy. After many years of studying the discrepancies, Kepler discovered that elliptical orbits with eccentricities close to that of a circle would fit the data quite nicely. This was the first of his three planetary laws which may be stated as follows:

1. A planet's orbit is an ellipse with the Sun at one focus.

2. The line from the Sun to any planet sweeps out equal areas of space in equal time intervals.

3. The revolution of a planet squared is proportional to the cube of its average distance from the Sun.

Kepler explained how the planets moved but still did not know why they moved. Newton would soon discover that gravity is the answer.

Before proceeding to Newton, a few words about Kepler's important laws are in order. Planets move in elliptical orbits. Roughly speaking, an ellipse is a flattened circle. The eccentricity of an ellipse is a measure of how flattened it is. However, the orbit of most of the planets in our solar system is very nearly a circle. If no great accuracy is required, calculation of a planet's position

Figure 1.11: *Isaac Newton co-invented the calculus (at the same time as Leibnitz) and discovered the Law of Universal Gravitation.*

can be greatly simplified assuming a circular orbit. From the second law, a consequence is that planets speed up as they approach the Sun and slow down as they move away from the Sun. Finally, due to the third law, it is possible to determine how long it takes a planet to revolve around the Sun (its "year") given its distance from the Sun and vice versa.

Sir Isaac Newton (1642–1727) stands out as one of the greatest geniuses of all time. He co-invented (with Leibnitz) a new branch of mathematics, called the calculus, which is a vital mathematical tool available to the sciences. His work completed the long scientific revolution which unseated the Aristotelian ideas that had dominated intellects for almost two thousand years. By necessity, however, the discussion here will have to be limited.

The Great Plague, which swept through Europe, caused universities to close their doors until the danger of disease was over. It was during this break from Cambridge University that Newton began his brilliant discoveries. Legend has it that while sitting under an apple tree, Newton saw an apple fall. This led him to conceive of a force that he called gravity. It is gravity, Newton reasoned, that accounts for the falling apple and for the motion of the planets in the solar system.

Stated more formally, Newton's *Law of Universal Gravitation* says that the force of attraction (gravity) between two objects is directly proportional to the product of the masses of the two objects and inversely proportional to the square of the distance between them. Several consequences are evident from this statement. First, the more massive the objects, the greater the attraction between them. So, since the Moon is less massive than the Earth, objects weigh less on the Moon's surface than they do on the Earth's surface. Secondly, objects accelerate as they fall towards the Earth because as the distance between two objects decreases, the force of attraction increases. This explains why planets speed up as they approach the Sun and slow down as they move away from it. Thirdly, all objects in the universe attract each other. However, due to small masses and large distances, the effects are often so small as to be unmeasurable. But for calculating planetary orbits precisely, it is necessary to account for the gravitational attraction between planets.

Newton's great work *Philosophiae Naturalis Principia Mathematica* was published in 1687. This monumental work demonstrated for the first time that the physical world can be explained and predicted by precise calculations. The science of dynamics, which tries to explain why objects move along a certain path, was thus born about 300 years ago. Newton, with his newly invented calculus, was able to show why the planets move in an elliptical orbit and, further, to show that comets may move in elliptical or parabolic orbits. Celestial motion was finally approachable by mathematical analysis.

Newton formulated three important laws which apply to all motion, celestial or otherwise. First, a body remains at rest or in uniform motion until acted upon by another force. This resistance to change in motion is called inertia and was actually discovered by Galileo. Newton's second law of motion relates force and acceleration while the first law relates mass and inertia. Mathematically, the second law can be written as

$$F = ma \tag{1.1.2}$$

where F is force, m is the object's mass, and a is the object's acceleration. From this equation it is easy to see that if an object's acceleration is zero, no external force is acting upon the object. The third law of motion states that for every action (force) there is an opposite and equal reaction. Rocket travel is possible because of the third law.

These three laws of motion plus the gravitation law were sufficient to accurately solve problems in dynamics for almost 300 years. But Newton's laws, however, could not adequately account for small, subtle effects in planetary motions. The explanation for such subtleties would require Einstein's *Theory of Relativity*. However, Newton's laws are good enough unless very high speeds

Figure 1.12: *Albert Einstein, Walter Adams, and W. W. Campbell at the 100-inch Telescope at Mt. Wilson Observatory in the 1930's. Associated Press Photo. Courtesy Mary Lea Shane Archives of the Lick Observatory.*

are involved just as Euclidean geometry is good enough until large distances are involved.

Albert Einstein (1879–1955) is often regarded as the greatest scientific genius of modern times. His *Theory of Relativity* shook the very foundations of classical physics and has reshaped our understanding of the universe. Einstein's theories are generally difficult for us to grasp. Perhaps we can understand how other generations felt when a Newton or Copernicus forced a reexamination of "common sense."

The most famous of Einstein's equations relates energy, mass, and the speed of light. Stated mathematically, we have

$$E = mc^2 \tag{1.1.3}$$

where E is energy, m is an object's mass, and c is the speed of light. A small quantity of mass can produce an incredible amount of energy. This conversion of mass to energy is the reason our Sun and distant stars produce great quantities of heat and light.

The most incredible consequence of Einstein's *Theory of Relativity* becomes apparent as an object's velocity approaches the speed of light. Let f be the value obtained from the equation

$$f = \sqrt{1 - \left(\frac{v^2}{c^2}\right)} \qquad (1.1.4)$$

where v represents the object's velocity and c is the speed of light. Then as the speed of light is approached, relativity theory predicts the following effects on length, time, and mass:

$$L' = Lf \text{ (Length Contraction)} \qquad (1.1.5)$$
$$T' = Tf \text{ (Time Dilation)} \qquad (1.1.6)$$
$$m' = m/f \text{ (Mass Increase).} \qquad (1.1.7)$$

In these equations L represents the object's length, T represents time, and m represents the object's mass. Note that if v is very small relative to c, the quantity under the square root is so close to 1 that the resulting change in L, T, or m is unmeasurable.

Einstein's equations have some incredible implications. As velocity approaches the speed of light (v approaches c), objects get smaller, are more massive, and time slows down for the object! For slow speeds the effect is so small that it is no wonder that Newton's equations have been so accurate and are still used except for the most exacting experiments. As an example, suppose an astronaut is traveling at 11.2 km/sec (7 miles/sec) which is the velocity required to escape the Earth's gravitational field. At this speed the astronaut's mass would increase by 0.0000001%, time for the astronaut would slow down by 0.00000007%, and the astronaut's length would shrink by 0.00000007%. Hardly detectable effects!

As velocities get closer to the speed of light, things get more interesting. Suppose our astronaut is 6 feet tall, weighs 200 pounds, and is to travel in his spaceship for 24 hours as measured by a clock on the Earth. (Actually, mass should be used here but the difference in mass and weight will not materially affect the point being made.) Table 1.1 demonstrates the changes that relativity theory predicts for different velocities. Velocities in the table are expressed as a percent of the speed of light. Note that weight increases rapidly as velocity approaches the speed of light but not nearly so rapidly for slower velocities.

From Table 1.1 it is easy to see why many scientists believe that achieving the speed of light is impossible. At the speed of light mass becomes infinite which means an infinite amount of energy is required to accelerate to the

Velocity	Weight	Time	Length
0.00	200.00	24.00	6.00
10.00	201.01	23.88	5.97
25.00	206.56	23.23	5.81
50.00	230.94	20.78	5.20
60.00	250.00	19.20	4.80
75.00	302.37	15.87	3.87
90.00	458.83	10.46	2.62
95.00	640.51	7.49	1.87
98.00	1,005.04	4.78	1.19
99.00	1,417.76	3.39	0.85
99.90	4,473.25	1.07	0.27
99.99	14,142.49	0.34	0.08

Table 1.1 Predictions From Relativity Theory. *From this table you can see that as velocity increases, so does weight but time and length become smaller!*

speed of light. If one could travel *faster* than the speed of light and if the theory of relativity holds, then one would be traveling *backwards* in time—but our astronaut would have shrunk to the size of a point!

Modern astronomers and physicists are trying to verify Einstein's remarkable theory. The search has led to the discovery of strange and wonderful new worlds within our universe. Of course Einstein could be wrong and our ideas of how the universe behaves could be as primitive and incorrect as the early Greeks. But the prospects of black holes, time travel, and parallel universes are tantalizing indeed! The universe may be even wilder than we can presently imagine.

1.2 Layout of the Book

This book begins with a discussion of some basic principles required for computing the location of astronomical objects and builds towards a climax at about Chapter 6. Chapter 2 shows how to perform conversions between several different units of measurement that are likely to be encountered in astronomy. Telescopes are discussed in Chapter 3. A telescope is the device most frequently used by amateur astronomers but selecting the proper telescope can be a confusing task. Amateurs are likely to buy the telescope with the highest power magnification that they can afford, but this is not always the desirable thing to do. Chapter 4 discusses time conversions which is probably the most difficult of all topics to understand. The conversion techniques are simple enough to perform but the reason for doing the conversions in the first place is not immediately obvious. In Chapter 5, the important topic of coordinate system conversions is discussed. Coordinate system conversions are necessary to account for the location of an observer with respect to the Earth and to account for changes in perspective such as viewing the solar system as geocentric rather than heliocentric.

The real fun in this book begins with Chapter 6. Chapter 6 combines the concepts from preceding chapters into a computer program which will calculate the location of a star for a given observer at a given date, time, and location. Furthermore, this program will produce a star chart for the date, time, and location in question. Chapters 7, 8, and 9 apply the same principles to predict the location of the Sun, Moon, and planets respectively. In addition, techniques are described for calculating all sorts of interesting things such as the phase of the Moon, distance to the planets, time of Sunrise and Sunset, and weight of an observer on different planets. Chapter 10 completes the book with a discussion on how to use an ephemeris as well as other astronomical aids.

The appendices contain helpful information related to the text. Appendix A contains a definition of astronomical terms used in the text as well as some terms that may be encountered in other astronomy books. Appendix B contains a list of mathematical symbols used in the various formulas. For those wishing to learn more about astronomy, Appendix C contains a selected list of recommended references.

A diskette containing each of the example programs is also included with this book. Calculations presented in the example programs are generally shown to 6 significant digits for intermediate calculations while final results are normally shown only to 2 decimal places. The reason for the extended precision is to allow one to compare calculations made by hand to the computer's calculations at each step so as to be able to detect where an error might have been made by the reader. The reader should be aware that the calculations are not generally correct to 6 decimal places and hence may differ slightly from calculations done by hand.

If one should compare results obtained from this book with published results in star charts or astronomy journals, there will likely be discrepancies. There are several reasons for such discrepancies. Differences are most likely to occur because of roundoff errors and because approximate formulas are used instead of exact formulas. However, the results should be more than accurate enough for general use. After all, for most amateur uses, a result accurate to a few minutes of arc is probably more than adequate.

In addition to roundoff errors and approximation formulas, errors can be introduced due to the observer's location. That is, if the observer only has an approximate longitude and latitude for his location, the results produced by the programs may be in error. Furthermore, the programs and techniques contained in this book are not intended to work for all possible dates. While all of the procedures should work for years in a range from about 1800 AD to 2100 AD, not all procedures will work for years outside that range.

Some readers may wish to use a hand calculator to follow along with the text or may wish to modify the programs for another application or another computer. Several points of caution are in order.

1. The formulas and examples used in this book are not necessarily accurate to as many decimal places as are printed out by the computer programs. Six decimal places are printed out to guide you in comparing your computations with those of the computer, not as an indication of a high degree of accuracy.

2. Beware of calculators which return incorrect results for small angles.

3. Beware of inverse trigonometric functions. Computers and calculators usually return a result that is between plus or minus 90° but it is generally necessary to adjust the result to be sure the answer is in the correct quadrant. This will be explained more fully in Chapter 5.

4. Beware of the difference between `FIX` and `INT` on calculators and computers. The problem only surfaces when negative numbers are considered. Throughout this book, `FIX` will be a function which returns the integer part of the argument while `INT` will be a function which returns the integer that is less than or equal to the argument. Some examples will illustrate the difference.

   INT(1.5) = 1
   INT(1.4) = 1
   INT(−1.5) = −2
   INT(−1.4) = −2
   FIX(1.5) = 1
   FIX(1.4) = 1
   FIX(−1.5) = −1
   FIX(−1.4) = −1

1.3 About the Program

The first sample program is entitled BLDSTARS. This program can be used to build data files containing star names and locations. Chapter 6 will make use of such data files to produce star charts. Several stars from *Sky Catalogue 2000.0* by Hirshfeld and Sinnott have already been entered into the included data file STARS.DAT, but you may use BLDSTARS to add additional celestial objects or create other star data files. The included file gives the star location as well as the *Smithsonian Astrophysical Observatory Star Catalog* designation.

It is worth taking a moment to examine BLDSTARS to see how it operates. All programs in this book are menu driven in exactly the same manner as BLDSTARS.

The procedure described here for running a program is the same for all of the included programs and will be assumed for the rest of the book.

To start BLDSTARS, start up BASIC as you normally would. Refer to your BASIC manual if you need directions. If you have a hard disk and have copied the programs onto your hard disk, use the CHDIR command to change to the directory containing the astronomy programs. If you have a floppy based system or have not copied the programs onto a hard disk, place the floppy diskette in disk drive A and type

```
CHDIR A:
```

and then press the RETURN key. This places you in the proper directory for executing the astronomy programs.

Once you have started BASIC and are in the proper directory, type

```
RUN "BLDSTARS"
```

and then press the RETURN key. This will cause the program to be loaded from the disk into memory and to start running. When BLDSTARS starts, the screen will clear and a menu of options will appear on the screen. Selection of a menu entry is accomplished by using the up and down arrow keys on the keyboard to highlight the desired option. Once the desired option is highlighted, pressing the RETURN key selects the option.

Chapter 2

Unit Conversions

The ability to convert from one unit of measurement to another is frequently necessary in astronomy and other physical sciences. This is true for a variety of reasons which will become apparent as the discussion progresses through the chapters ahead. This chapter will cover the unit conversions that are used throughout the rest of the book as well as some other conversions that are frequently encountered in basic astronomy.

Many unit conversions arise from a need to convert from one system of measurements to another as in converting from the English to the metric system. Sometimes conversions are required within a measurement system to convert to more convenient units. For example, we often convert between inches and feet or feet and miles so as to express a measurement in more convenient units.

The English system (inch, pound, gallon, etc.) is rarely used in science. At least one reason for this is that conversions within the English system are more complicated than necessary. Consider measuring length in the English system. The smallest unit is the inch. Inches are converted to feet by dividing by 12, feet are converted to yards by dividing by 3, and feet are converted to miles by dividing by 5,280. There is no apparent rhyme nor reason for these conversion factors. An additional problem is that the words used to indicate length give no clue as to how to perform the conversion. This confusion in the English system has arisen for historical reasons as the measuring units were developed.

The metric system (meter, gram, liter, etc.) is orderly and consistent. All conversions in the metric system are done by simply multiplying or dividing by an appropriate power of 10. This has the effect of merely moving a decimal point which can be done mentally without the need for actually doing a multiplication or division. For example, a meter is the basic metric unit for measuring length. Meters are converted to centimeters (the prefix *centi* means one hundreth) by dividing by 0.01 (move the decimal point 2 places to the right). Meters are converted to kilometers (the prefix *kilo* means one thousand) by dividing by 1,000 (move the decimal point 3 places to the left). Thus, 1 meter equals 100 centimeters while 5,000 meters equals 5 kilometers. The conversion factor to use is immediately obvious in the metric system because of the word's prefix. Table 2.1 lists the factors for some of the commonly used prefixes in the metric system.

Prefix	Factor
micro	0.000001
milli	0.001
centi	0.01
kilo	1,000
mega	1,000,000

Table 2.1 Metric System Prefixes. *In the metric system, prefixes equate to conversion factors.*

2.1 Distance Conversions

Distance conversions can be accomplished quite easily by making use of a simple mathematical trick called crossmultiplying. The essence of the trick is that if

$$\frac{a}{b} = \frac{c}{d} \tag{2.1.1}$$

then

$$ad = cb. \tag{2.1.2}$$

Equation 2.1.1 is arrived at by taking the original equation and multiplying the numerator from each side of the equation with the denominator from the other side of the equation. Since multiplication is commutative, the order in which a, b, c, and d are multiplied is unimportant. Thus, equation 2.1.1 can also be written as

$$ad = bc. \tag{2.1.3}$$

To illustrate the concept of crossmultiplication, consider converting from inches to feet. Specifically, convert 4 feet to the equivalent number of inches. The first step is to form an equation involving fractions which represents the problem. Since it is known that 1 foot equals 12 inches, we write

$$1 \text{ foot} = 12 \text{ inches}.$$

Next, let the variable x represent the quantity we are looking for. Write the known quantity, 4 feet, underneath the left hand side of the equation above and the unknown quantity, x, underneath the right hand side of the equation above. This equation forms a proportional ratio that relates feet and inches. Expressed mathematically, we have

$$\frac{1 \text{ foot}}{4 \text{ feet}} = \frac{12 \text{ inches}}{x}.$$

The key to forming the fraction correctly is to make sure all the like units are on the same side of the equation. Thus,

$$\frac{1 \text{ foot}}{x} = \frac{12 \text{ inches}}{4 \text{ feet}}$$

is incorrect because like units are not on the same side of the equation.

After forming a correct fraction which represents the problem, crossmultiply by applying equation 2.1.1 where $a = 1$ foot, $b = 4$ feet, $c = 12$ inches, and $d = x$. Thus, we have

$$(1 \text{ foot})(x) = (12 \text{ inches})(4 \text{ feet}).$$

Now, isolate the unknown by itself by dividing both sides of the equation by 1 foot. This gives,

$$x = \frac{(12 \text{ inches})(4 \text{ feet})}{1 \text{ foot}}.$$

The units "feet" and "foot" cancel out leaving only "inches" on the right hand side of the equation. This is another check that the problem has been formulated correctly because we expect the result to be in inches. The equation now looks like

$$x = \frac{(12 \text{ inches})(4)}{1}.$$

The final step is to simply complete the arithmetic operations in the equation. In this case, multiply 12 by 4 and divide by 1. This gives

$$x = 48 \text{ inches}$$

which is the correct result.

When properly done, crossmultiplication is a quick and easy method for doing conversions. It has the advantage that one only has to memorize a single equation which expresses the relationship between the different units of measurement. In the example above, it is possible to convert inches to feet or feet to inches by simply knowing that 1 foot equals 12 inches. There is no need to also memorize that 1 inch is about 0.083 feet although this second relation would work equally well.

Unit of Time	Equivalent Angle
1 day	360°
1^h	15°
1^m	15′
1^s	15″

Table 2.2 Converting Time to Angles *Time can be measured in degrees by noting that the Earth rotates 360° in 24 hours.*

2.2 Applying Crossmultiplication

The crossmultiplication technique is not restricted to just distance conversions. This technique will work for any conversion in which there is a simple relationship between the two units that involves only a multiplication or division. For example, it is common practice in astronomy to convert time to degrees. Since the Earth rotates in 24 hours, there is a simple relationship between time and degrees. That is, 24 hours equals 360°.

Consider converting 5 hours to the equivalent number of degrees. First, we write down the fraction representing the problem following the rules outlined above. The correct fraction is

$$\frac{24 \text{ hours}}{5 \text{ hours}} = \frac{360°}{x}.$$

Crossmultiplying, using equation 2.1.2, yields

$$(24 \text{ hours})(x) = (360°)(5 \text{ hours}).$$

Dividing both sides by 24 hours and canceling out the "hours" units gives

$$x = \frac{(360°)(5)}{24}$$

which reduces to

$$x = 75°.$$

Note that degrees can be subdivided into minutes and seconds which are also units for measuring time. When there might be confusion as to which is meant, the terms arcseconds and arcminutes are used to distinguish between time and angles. Tables 2.2 and 2.3 contain the proper conversion equations relating time and angles. The formulas in Table 2.2 are exact while some of the ones in Table 2.3 have been rounded. The program described at the end of this chapter implements the crossmultiplication technique for the formulas found in Table 2.4.

2.3 Scientific Notation

Astronomers and other scientists must frequently deal with measurements that range from very small to very

Angle	Equivalent Time
1 radian	3.82^h
1°	4^m
1′	4^s
1″	0.067^s

Table 2.3 Converting Angles to Time. *Time conversions are a prime calculation in determining where an object appears for an observer.*

25.4 mm	1 inch
0.3048 m	1 foot
1.609344 km	1 mile
1 light year	5.87E12 miles
1 light year	0.3068 parsecs
1 A.U.	9.29E7 miles
180°	3.14159 radians
360°	24^h

Table 2.4 Conversion Formulas. *Astronomy often requires measurements ranging from the very small to the very large.*

large. For example, astronomers often measure the wavelength of light reaching Earth from distant stars to determine the materials that make up the stars. A typical wavelength measurement is on the order of 1 one hundred millionth of a centimeter (0.00000001 cm). At the other extreme, astronomers measure distances that are so vast that they often exceed trillions of miles (1 trillion miles is 1,000,000,000,000 miles). It is both inconvenient and error prone to write down so many zeroes to express numbers such as these. (Inadvertently dropping three zeroes changes a trillion miles to a mere billion miles.) For this reason, scientists normally express numbers in scientific notation.

A number expressed in scientific notation is written as the product of two numbers. The first number is between 1 and 10, that is, a number with a single non-zero digit to the left of the decimal point. The second number in the product is a scaling factor, written as a power of ten, which indicates where to place the decimal point. For example, 9.3×10^7 is the proper way to express the average distance from the Sun to the Earth in scientific notation. Since

$$10^7 = 10,000,000$$

we have

$$9.3 \times 10^7 = 9.3 \times 10,000,000 = 93,000,000.$$

As another example, consider 1.5×10^{-4}. Since

$$10^{-4} = 0.0001,$$

we have

$$1.5 \times 10^{-4} = 1.5 \times 0.0001 = 0.00015.$$

The easiest way to deal with scientific notation is to remember that the power of 10 exponent indicates how many digits to the left (for negative exponents) or to the right (for positive exponents) to place the decimal point.

Table 2.4 contains several units and notations that may be unfamiliar. A number with the letter "E" in it represents a number in scientific notation. For example, 9.29E7 stands for the number 9.29×10^7. Expressing numbers in scientific notation by using the letter "E" will be used throughout the book to conform to common computer usage and to match the computer program printouts.

As has been pointed out, astronomical distances can be very large numbers when measured in miles or kilometers. Even when expressed in scientific notation, such measurements are unwieldly to manipulate. Therefore, astronomers have defined other measurements when distances are so large. For distances within the solar system, the Astronomical Unit (A.U.) is normally used. One A.U. is the distance from the Earth to the Sun. Thus, the distance from the Earth to the Sun would normally be expressed as 1.0 A.U. instead of 9.3×10^7 miles. When distances are as far away as the stars, light years are the units normally used. A light year is pretty much what the term implies. A light year is the distance that light travels in one year. Another unit, the parsec, is sometimes used to measure distances that are of the same magnitude as light years. A parsec is approximately 3.26 light years.

2.4 Temperature Conversions

Temperature measurements in the United States are normally expressed as degrees Fahrenheit. Scientific measurements, however, are normally expressed as degrees Celsius. (Scientific degrees may also be expressed in degrees centigrade which means the same thing as degrees Celsius.) In this section, formulas for converting between the two systems will be described for completeness sake even though temperature conversions will not be required throughout the remainder of the book.

Temperature conversion, although simple, is not as straightforward to remember as the crossmultiplication technique given above. The required equations are

$$C^\circ = \frac{5}{9}(F^\circ - 32) \tag{2.4.1}$$

and

$$F^\circ = 32 + \frac{9}{5}C^\circ. \tag{2.4.2}$$

where F represents degrees Fahrenheit and C represents degrees Celsius.

2.5 Decimal Format Conversions

It is common practice to express time in terms of hours, minutes, and seconds. One might, for example, say that the time is 4 hours, 32 minutes, and 29 seconds Central Standard Time. This format is called the HMS format and is written as $4^h32^m29^s$. (Note: The computer programs will indicate HMS time as 4h 32m 29s since superscripts are difficult to produce.) To avoid difficulties with whether the time meant is a.m. or p.m., a 24 hour clock will be assumed throughout the text. Thus, 1:30 p.m. is expressed in HMS format as $13^h30^m0^s$. Angles are often expressed in a similar format called DMS format and are written as $24°13'18''$. (The computer programs will indicate DMS format as 24deg 13m 18s).

The HMS and DMS formats are not very convenient for computational purposes. Angles and time are converted to a decimal format first so that only one measurement unit is involved instead of three. Once in the decimal format, arithmetic computations such as addition and subtraction are much easier to perform since there is no need to consider adding the units separately. For instance, the decimal format for $4^h30^m0^s$ is 4.5^h since 30 minutes is 0.5 hours. Adding 1.5^h ($1^h30^m0^s$) to 4.5^h gives 6^h without the need to consider the minutes and seconds separately.

Note that $4^h30^m0^s$ is not the same as 4.300^h nor is the time 12:30 the same as 12.30^h since 0.3 hours is 18 minutes, not 30. Also make note of the fact that the ranges for numbers in the decimal format depend upon whether the number represents angles or represents time. When expressed in decimal format, time is in the range $[0^h, 24^h]$ while angles are in the range $[-360°, 360°]$. Thus, an angle can be a negative number but time is always positive.

The procedure for converting the HMS format to the decimal format is the same as for converting DMS format to decimal format. Likewise, converting back from decimal to HMS format is identical to converting decimal format to DMS. Dealing with angles is slightly more complicated because the angle can be negative and care must be taken to account for the sign of the number. The conversion procedures below will deal only with angles. Everywhere degrees are mentioned below, substitute hours and the procedure will work for time conversions as well.

Converting DMS to decimal format requires 7 steps. Assume that *Degrees*, *Minutes*, and *Seconds* are the variables containing the DMS format of the angle. The procedure is as follows:

1. If $Degrees < 0$ let $Sign = -1$ else $Sign = 1$. Note that *Sign* is 1 if *Degrees* is 0. Many computers and calculators have a function which returns the sign of a number, but normally return a result of zero when the number is zero.

2. $Degrees = |Degrees|$ where $|x|$ means to take the absolute value of x.

3. Convert the seconds to decimal minutes. This is done by the equation:
$$dm = \frac{Seconds}{60}.$$

4. Add step 3 to *Minutes* to obtain total minutes. Thus
$$Total\ Minutes = dm + Minutes.$$

5. Convert total minutes to decimal degrees by
$$dec = \frac{Total\ Minutes}{60}.$$

6. Add the results of step 5 to *Degrees*.
$$Degrees = dec + Degrees.$$

7. Finally, account for the possibility that the angle was negative. Thus,
$$Decimal\ Degrees = Sign * (Degrees \text{ from step 6}).$$

The "*" in step 7 is the conventional way that computer languages represent the multiplication operation and will be used frequently in this book. If dealing with time, the procedure above can be simplified by omitting steps 1, 2, and 7 since there is no need to be concerned about the possibility of a negative number.

Converting from decimal format to DMS format requires 6 steps. Assume that the variable *Dec* contains the angle in decimal format. Furthermore, assume that the function `FRAC` returns the fractional part of a number. Thus, $\texttt{FRAC}(1.5) = 0.5$ and $\texttt{FRAC}(-1.5) = 0.5$. The required steps are:

1. If $Dec < 0$ let $Sign = -1$ else $Sign = 1$.

2. $Dec = |Dec|$.

3. $Degrees = \texttt{INT}(Dec)$.

4. $Minutes = \texttt{INT}(60 * \texttt{FRAC}(Dec))$ since $\texttt{FRAC}(Dec)$ gives the total number of minutes expressed as a fractional part of the degrees and multiplying by 60 converts this to minutes.

5. $Seconds = 60 * \texttt{FRAC}[60 * \texttt{FRAC}(Dec)]$.

6. $Degrees = Sign * Degrees$. to account for the possibility of a negative angle.

If the desired conversion is to convert time to HMS format, steps 1, 2, and 6 can be omitted.

Both conversion procedures described above can be tricky to implement due to the need to consider negative angles. The problem comes in when the angle is negative but greater than $-1°$ as in $-0.586°$. The program for this chapter, UNITCON, automatically handles negative angles that are greater than $-1°$.

2.6 Data Accuracy

Before continuing with the chapters ahead, an important concept to discuss is accuracy. Accuracy is a statement of how well a measurement agrees with the actual or accepted value of the entity being measured. For example, suppose a ruler with divisions every inch is used to measure the length of a book and the result is 8 inches. This number is really only an approximation of the book's true length. In actuality, is the book closer to 8.1 inches in length, 7.9 inches, or 8.02 inches? A better measurement could be obtained by using a ruler that has divisions every tenth of an inch. With such a ruler, the book's length could be measured to the closest tenth of an inch rather than to the closest inch. Similarly, a ruler could be constructed that has even finer divisions, say every hundredth of an inch, to give an even better measure of the length. However, it is easily seen that physical considerations place a limit on how finely a ruler may be sudivided.

Any measurement of a physical entity (length, time, temperature, etc.) is by necessity an approximation. Astonomy, and all the sciences, use all kinds of measurements (e.g., a star's location, the instant a planet passes through a node, the distance to the Sun) that must be understood as approximations. This is not to say that all numbers encountered are approximations. The length of a mile is exactly 5,280 feet by definition. By definition, there are exactly 1,000 meters in a kilometer. These are not approximations because they are defined to be exact as opposed to being the result of some measurement. However, measurements which use these definitions must still be viewed as approximations.

There are several reasons that approximations arise when measurements are made. First, as already demonstrated, the accuracy of the instrument used places a limit on the accuracy of the measurement. Second, a human must typically judge how close a measurement falls within the limits of his instrument. (When measuring with a ruler, what should one do when the length falls somewhere between two divisions?) Third, approximate formulas may be used to derive another measurement. For example, the chapter on locating the position of a planet uses an approximation to find the solution to Kepler's equation. This approximation is in turn used in another formula to estimate the position of a particular planet. Hence, the final solution, based upon approximate measurements and an approximation formula, will not give the exact location of a planet.

Why is the question of accuracy important? Because results obtained from approximations may imply an accuracy greater than is really the case. As an example, suppose a ruler subdivided every tenth of an inch is to be used to find the area enclosed by a rectangle. Assume that the length is measured to be 3.1 inches and the width 0.5 inches. The area is calculated by multiplying these two numbers to give

$$\text{Area} = (3.1 \text{ inches})(0.5 \text{ inches}) = 1.55 \text{ Square Inches}.$$

This result implies that the area is known to the nearest hundredth of an inch when our original measurements were known only to the nearest tenth of an inch! In the context of the algorithms used in the chapters ahead, a calculation might imply that an event, such as Sunrise, can be computed to the nearest second when in reality the result may be accurate only to the nearest 5 minutes. The situation is compounded because a computer can easily perform calculations with many more digits of accuracy than the original measurements warrant.

The convention normally used in scientific measurements to specify the accuracy of a measurement is to give all the accurate digits plus a single digit of uncertainty. For instance, if a ruler is graduated in millimeter increments and the measurement stated is 21.3 millimeters, there are three significant digits. The digit 3 is a digit of uncertainty because an estimate had to be made of where the actual length was between 21.0 and 22.0 millimeters.

When performing arithmetic operations on numbers, the result can not be more accurate than the least accurate measurement. As an example, suppose the two numbers 21.5 and 0.003 are to be multiplied together. Since the least accurate measurement is 21.5 (its accuracy is known only to the first decimal place whereas 0.003 is known to 3 decimal places), the resulting product will have only one significant digit to the right of the decimal point. Thus,

$$(21.5)(0.003) = 0.0645$$

which must be rounded to the nearest tenth giving an approximate answer of 0.1.

In writing this book, it was difficult to strike a balance between always giving the correct number of significant digits and in providing enough digits so that readers could compare their calculations with the examples given. The approach that will be taken for the remainder of the book, and in the computer programs, is to give all calculations to 6 decimal places. This *does not* imply an accuracy of 6 significant digits. In general, the algorithms used here are correct to only a few minutes of arc or a few minutes of time.

2.7 About the Program

The program UNITCON demonstrates the conversion techniques described in this chapter. UNITCON is started just like BLDSTARS described in Chapter 1. The menu entries in this program perform conversions from one system of measurements to another. Typing the RETURN key in response to a request for an input value indicates to the program to which units you wish to convert. For example, suppose you want to do conversions between meters and feet. UNITCON will ask you for a value in meters and for a value in feet. To convert from meters to feet, type a RETURN when asked for meters and type in the desired number of feet when requested to do so. To convert from feet to meters, do just the opposite. Type in the number of meters when requested to do so, and type RETURN when prompted for the number of feet.

The program UNITCON normally performs its calculations without displaying any intermediate results. The second menu entry toggles between displaying intermediate results at each step and suppressing them. If you want to see the intermediate calculations, select this menu entry before beginning a calculation. Once selected, all intermediate steps will be displayed until you select the second menu entry again to turn off the display of intermediate steps.

Each of the chapters ahead follow the same convention as UNITCON. When the intermediate results are displayed, they will correspond to the steps in the algorithms discussed to make it easier to compare your manual calculations with those of the computer.

2.8 Sample Problems

Some sample problems for this chapter are listed here along with the correct answers. The results may differ slightly on a different computer or on a calculator from what is presented here. If the results do not agree very closely, check your arithmetic to see if an error has been made. Practice doing conversions both ways such as converting a number from miles to kilometers and then converting it back from kilometers to miles.

1. Convert 5mm to inches.
   (Ans: 0.196850 inches)
2. Convert 10 inches to mm.
   (Ans: 254mm)
3. Convert 30 meters to feet.
   (Ans: 98.42519 feet)
4. Convert 25 feet to meters.
   (Ans: 7.62 meters)
5. Convert 100 miles to kilometers.
   (Ans: 160.9344 km)
6. Convert 88 km to miles.
   (Ans: 54.68067 miles)
7. Convert 12 light years to miles.
   (Ans: 7.044E13 miles)
8. Convert 9.3E7 miles to light years.
   (Ans: 1.584327E-5 light years)
9. Convert 5 light years to parsecs.
   (Ans: 1.534 parsecs)
10. Convert 3 parsecs to light years.
    (Ans: 9.778356 light years)
11. Convert 2 AU to miles.
    (Ans: 1.858E8 miles)
12. Convert 10,000 miles to AU.
    (Ans: 1.076426E-4 AU)
13. Convert 180° to radians.
    (Ans: 3.14159 radians)
14. Convert 2.5 radians to degrees.
    (Ans: 143.2396°)
15. Convert $10^h25^m11^s$ to decimal hours.
    (Ans: 10.419722^h)
16. Convert 20.352^h to HMS format.
    (Ans: $20^h21^m7.2^s$)
17. Convert 2^h to degrees.
    (Ans: 30.0°)
18. Convert 156.3° to hours.
    (Ans: 10.42^h)
19. Convert −13°4′10″ to decimal degrees.
    (Ans: −13.069444°)
20. Convert −0.508333° to DMS format.
    (Ans: −0°30′30″)
21. Convert 300°20′0″ to decimal degrees.
    (Ans: 300.333333°)
22. Convert 10.2958° to DMS format.
    (Ans: 10°17′44.88″)
23. Convert 100° C to F°.
    (Ans: 212° F)
24. Convert 32° F to C°.
    (Ans: 0° C)

Chapter 3

Telescopes

Astronomy is a visual science. The beauty of a moonlit night, the magical appearance of a shooting star, and the wispy strands of the Milky Way are appealing sights indeed. Perhaps it is this ethereal beauty that first entices and excites interest in astronomy. Even with the naked eye, the universe is an enchanting and thrilling wonderland.

Besides appreciating nighttime splendors, the amateur can do a lot with just the naked eye. As an example, learning the constellations is a rewarding experience which requires nothing more than a willingness to be observant. The naked eye is the most basic, and cheapest, instrument for astronomical study. Until Galileo's time this was about all that was available. However, with even a moderately sized telescope the universe becomes a panorama of galaxies, twinkling stars, colorful nebulae, and mysterious planets. Because of the greatly increased view of the universe, telescopes are an important tool for amateur and professional alike.

Selecting the proper telescope can be a bewildering task. There are many different types and sizes in various price ranges. Many select a telescope by choosing the one with the highest magnification power that fits within their budget. But a telescope's magnifying power is only one of several factors to consider in selecting a telescope and depending upon how the telescope is to be used, power may be one of the least significant factors. If the telescope is to be used primarily for studying lunar and planetary details, then magnification is an important factor. For observing deep space objects, magnification is not as important as light gathering ability since the distances involved are so vast.

As a general rule, choose light gathering ability (as measured by the size of the telescope's aperture) over magnifying power. In addition, one must consider how transportable the telescope is and must consider the size and cost of suitable mountings for large aperture telescopes. For example, as Jean Texereau points out in his book *How to Make a Telescope*, an 8-inch reflector with an altazimuth mount can be built with a total weight of 40 pounds or less and can be transported quite easily in an automobile. But a 20-inch telescope with a proper equatorial mount may weigh as much as 2 tons! (Other mountings, such as a Dobsonian, make large aperture telescopes portable but they have other disadvantages.) Such a telescope is hardly suitable for the casual weekend astronomer.

Figure 3.1: *This equatorial fork mount Newtonian telescope was displayed at Stellafane in 1988.*

A telescope's mounting is an important consideration that is often overlooked. A high quality telescope that is poorly mounted is virtually useless because the slightest vibration will cause the image to blur making observation impossible. Mountings are also an important consideration if it is desirable to take photographs. To take photographs, the mounting should have a clock drive attached to it so that the telescope will track the motion of the object during the time—which can range from several seconds to an hour—that the camera shutter is open. An equatorial mount with a clock drive is virtually a necessity for astrophotography.

A detailed and rigorous discussion of optics, telescopes, and mountings is beyond the scope of this book. However, for completeness sake, this chapter contains a brief overview of telescope optics, practical limitations, types of telescopes and mountings, and eyepieces. More information can be found in several of the works cited in Appendix C.

A few terms need to be defined before proceeding to the following sections. A telescope's objective is the lens or mirror that gathers the light rays. The diameter of the objective is called the aperture. If the objective is a mirror, the telescope is called a reflector whereas if the objective is a lens, the telescope is called a refractor. A more detailed discussion of telescope construction will be given below, but it will be useful to describe these terms here for use in subsequent discussion.

3.1 Fundamental Optics

This section gives a very brief introduction to the highly technical area of optics. The more one understands about how telescope optics work, the more one is able to wisely select and use a telescope. The discussion which follows will describe various properties of lenses, but the same discussion generally applies to the concave mirrors used in reflector telescopes. Differences between lenses and mirrors will be noted where appropriate. One important difference to keep in mind throughout the discussion is that lenses work by refraction, bending light rays as they pass through the glass, while mirrors work by reflection, bending light rays as they are reflected.

Light rays are bent whenever they pass from one medium, such as air, to another medium, such as glass or water. This is easily demonstrated by filling a glass with water and placing a pencil in the glass. The image of the pencil when viewed through the water is enlarged and appears to be bent with respect to the portion of the pencil that remains outside the water. Optical lenses work in the same way.

The reason light rays are bent when they pass from one medium to another is because the speed of the light ray changes. When passing from a less optically dense medium to a more optically dense medium, such as from air to glass, the light rays are slowed down. In the reverse situation, when light passes from an optically dense medium to a less optically dense medium, the light rays speed up. (The maximum speed which can be obtained, as explained in Chapter 1, is approximately 186,000 miles per second. This is the speed of light in a vacuum.) In both cases, the light rays are bent in passing from one medium to another.

The amount by which light rays are bent depends upon the medium that they are passing through. Light

Material	n
Vacuum	1.00
Air	1.003
Water	1.33
Ethanol	1.36
Fluorite	1.43
Fused Quartz	1.46
Lucite or Plexiglas	1.51
Crown Glass	1.52
Flint Glass	1.66
Diamond	2.42

Table 3.1: Refractive Indices. *Different substances bend light rays by different amounts.*

rays are bent much less when entering air than they are when entering a more solid substance such as glass. The refractive index of a material is a measurement that describes how much light rays will be bent as they pass through a substance. Mathematically, the refractive index n is the ratio of the speed of light in a vacuum to the speed of light in the substance. That is,

$$n = \frac{\text{Speed of light in vacuum}}{\text{Speed of light in substance}}. \tag{3.1.1}$$

Table 3.1 gives some typical refractive index values for various substances. Note that since air does not have a refractive index of 1.0, light rays are bent when they enter the Earth's atmosphere from space. This means that the apparent position of a star (e.g., where we "see" it) is different from the actual position of the star. Corrections to account for the refraction of the Earth's atmosphere will be discussed later in Chapter 5.

A typical telescope lens arrangement is shown in Figure 3.2. This lens is a compound lens made up of a concave and a convex lens surface. Light rays entering at the top of the lens are bent downwards toward the lens center while light rays entering from the bottom are bent upwards toward the lens center. In a perfect lens, all the light rays will converge at the focal point. The distance from the center of the lens to the focal point is called the focal length. As will be explained shortly, the focal length of a lens plays an important role in determining how much an image is magnified.

Lenses such as the one in Figure 3.2 are called converging lenses. Any lens which is thicker in the middle than at the ends is a converging lens and has the property that light rays will converge at a single point, the focal point. A lens which is thinner in the middle than at the ends, as depicted in Figure 3.3, is called a diverging lens. Such lenses have a focal point as well, but the focal point is the point from which the light rays appear to *diverge* as opposed to *converge* in the case of converging lenses.

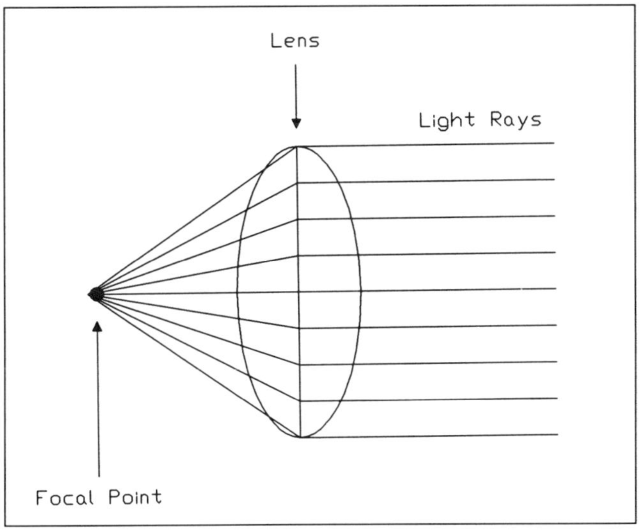

Figure 3.2: *This type of lens is a converging lens. All of the light rays passing through the lens converge at a single point called the focal point.*

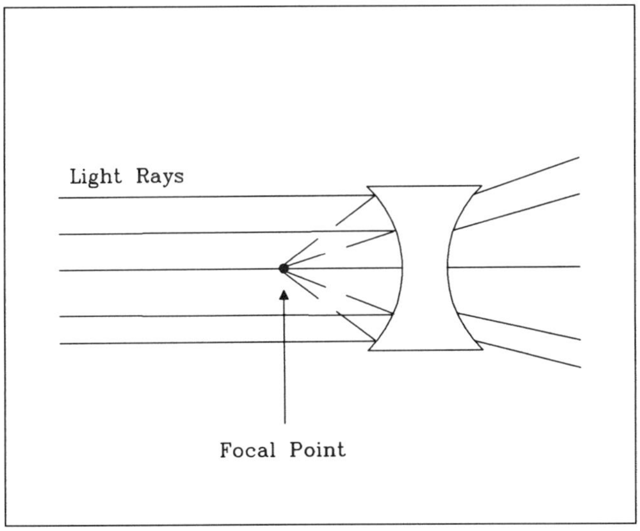

Figure 3.3: *Diverging lenses refract light rays away from the center of the lens. They are thinner in the middle than at the edges of the lens.*

Mirrors can be constructed to achieve the same results as converging and diverging lenses. However, the physics of *how* they achieve the results are different from lenses. Lenses operate by refracting light, that is, causing light rays to bend based upon a change in the light ray's speed as it passes from one optical medium to another. Mirrors work on the principle of reflection, that is, changing the direction of a light ray by reflection, not by changing the light ray's speed. When light rays strike a reflective surface at an angle, they are reflected from

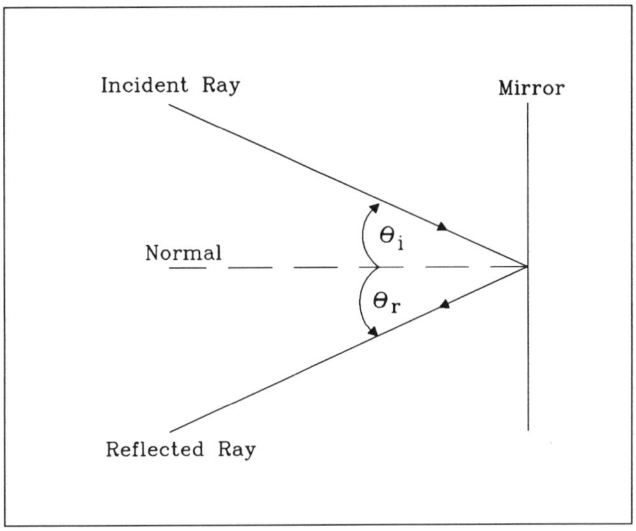

Figure 3.4: *Mirrors change the direction of a light ray by reflecting the light ray. The angle of incidence is always equal to the angle of reflection.*

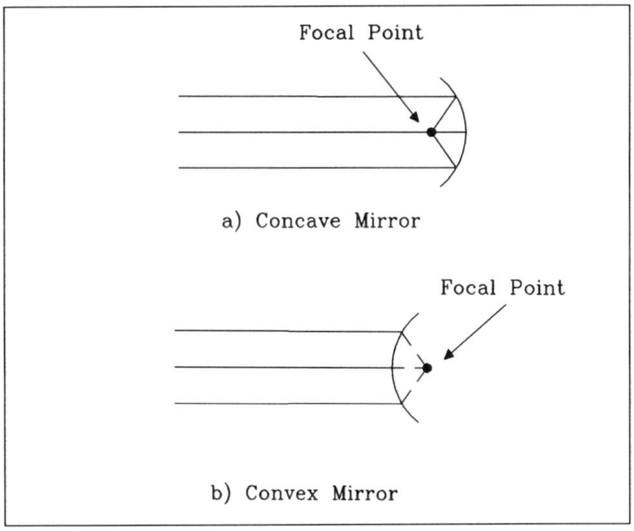

Figure 3.5: *Concave mirrors are formed when the reflective surface is inside a portion of a sphere or parabola. Convex mirrors are formed when the reflective surface is outside the sphere or parabola.*

the surface at an angle that is equal to the original angle. Figure 3.4 illustrates the situation. Angle θ_i is called the angle of incidence and is equal to angle θ_r, the angle of reflection.

If a mirror is spherical or parabolic in shape, it will have a focal point just as a lens does. A concave mirror, as shown in Figure 3.5, reflects light rays so that they are inside the center of curvature of the mirror. This situation is analogous to a converging lens.

A convex mirror, also shown in Figure 3.5, reflects light rays away from the curve of the mirror. This situation is analogous to a diverging lens and explains why concave mirrors are used for reflector telescopes instead of convex mirrors. The equations which follow are described for a lens, but unless otherwise noted, they apply to mirrors as well.

A term closely associated with the focal length of a lens (or mirror) is the focal ratio. The focal ratio is determined by dividing the lens's focal length by the diameter of the lens. Mathematically, we have

$$f \text{ ratio} = \frac{f}{D} \qquad (3.1.2)$$

where D is the diameter of the lens and f is the focal length of the lens. The focal ratio, sometimes called the f number, is typically written as $f/4.5$ to indicate that the ratio of the focal length to the aperture size is 4.5. Note that the focal length and lens diameter must be expressed in the same units, typically millimeters, to use equation 3.1.2.

Using equation 3.1.2, we find that a 4-inch (101.6mm) telescope with an objective lens focal length of 200mm is

$$f \text{ ratio} = \frac{200\text{mm}}{101.6\text{mm}} \approx 2.$$

However, a 6-inch (152.4mm) telescope with an objective lens focal length of 300mm also has an f number of $f/2$. Thus, the focal ratio alone is not sufficient for distinguishing one telescope from another.

The focal length for a lens depends upon the curvature of the lens and the refractive index of the lens. Shorter focal lengths are produced by more curvature and longer focal lengths are produced by less curvature. Refering to Figure 3.2, let R_1 be the radius of curvature for the convex lens (the part of the lens closest to the light source) and let R_2 be the radius of curvature for the concave lens (the part of the lens closest to the focal point). Given n, the refractive index of the material used in the lens, the focal length f is given by the equation

$$\frac{1}{f} = (n-1)\left(\frac{1}{R_1} - \frac{1}{R_2}\right) \qquad (3.1.3)$$

As an example, assume that the lens in Figure 3.2 has a refractive index of $n = 1.5$, $R_1 = 20$mm, and $R_2 = -80$mm. (Note that the radius of curvature for a concave lens is given as a negative number.) Applying equation 3.1.3, we have

$$\begin{aligned} \frac{1}{f} &= (1.5-1)\left(\frac{1}{20} - \frac{1}{-80}\right) \\ &= (0.5)(0.05 - (-0.0125)) \\ &= (0.5)(0.05 + 0.0125) \\ &= (0.5)(0.0625) \\ &= 0.03125. \end{aligned}$$

or

$$\frac{1}{f} = \frac{0.03125}{1}.$$

Using the cross-multiplication technique from Chapter 2 to solve for f, we have

$$f = \frac{1}{0.03125} = 32\text{mm}.$$

A concave mirror, used in reflector telescopes, also has a focal point and focal length but calculating the focal length does not involve an index of refraction. The calculation of the focal length, assuming 100% reflectivity and a radius curvature of R is given by

$$f = \frac{-R}{2}. \qquad (3.1.4)$$

(Keep in mind that the radius curvature for a concave surface is given as a negative number.) Assuming a concave mirror has a radius curvature $R = -100$mm, its focal length is

$$f = \frac{-(-100)}{2} = \frac{100}{2} = 50\text{mm}.$$

When an object is viewed through a lens, it is inverted and appears "upside down." The reason for this is demonstrated in Figure 3.6. Light rays from the top of the object being viewed are bent downward while those from the bottom are bent upward. When the light rays converge at the focal point (for objects at infinity) or beyond (for closer objects), the rays have crisscrossed and the result is an inverted image. This is why images in an astronomical telescope are inverted. By adding additional lenses, the image can be reinverted. Terrestial telescopes do precisely that and consist of at least three lenses: an objective lens, an eyepiece, and a lens to reinvert the image. Prisms are also commonly used to reinvert the image, especially in binoculars.

The magnifying power of a telescope can be calculated from the focal length of the objective lens (or objective mirror) and the focal length of the telescope's eyepiece. The magnifying power is given by the equation

$$Power = \frac{f_{obj}}{f_{eye}}. \qquad (3.1.5)$$

That is, simply divide the focal length of the objective lens by the focal length of the eyepiece. It is important to remember to be sure that both the focal length of the objective lens and the focal length of the eyepiece are expressed in the same units. Typically focal lengths are expressed in millimeters.

Equation 3.1.5 can be rewritten as

$$f_{eye} = \frac{f_{obj}}{Power}. \qquad (3.1.6)$$

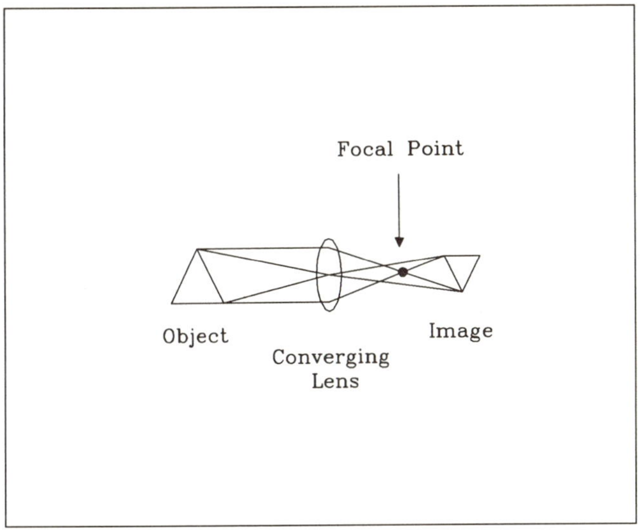

Figure 3.6: *Objects viewed through a lens are inverted.*

Expressed in this form, it is easy to calculate the eyepiece required to achieve a specified magnification power.

For example, suppose a telescope has an objective lens with a focal length of 100mm and two eyepieces with focal lengths of 20mm and 10mm respectively. Applying equation 3.1.5, we have

$$Power = \frac{100\text{ mm}}{20\text{ mm}} = 5$$

for the 20mm eyepiece and

$$Power = \frac{100\text{ mm}}{10\text{ mm}} = 10$$

for the 10mm eyepiece.

From equation 3.1.5 relating magnification and focal lengths, it is obvious that to achieve high magnification it is desirable for the focal length of the objective lens to be as *long* as possible while the focal length of the eyepiece should be as *short* as possible. Furthermore, to *increase* a telescope's magnifying power, *decrease* the focal length of the eyepiece. However, as we shall discuss in the next section, other factors place practical limits on a telescope's magnifying power and ultimate performance.

3.2 Practical Limitations

The discussion so far has assumed that optical materials, either for lenses or mirrors, are perfect and has ignored some important practical limitations that arise from properties of light. Lenses can not be made that are free from defects, but even if it were possible to make such lenses, physical properties of light place limitations on what can be achieved. In an ideal optical system consisting of a converging lens (or mirror), all light rays from an infinite distance will precisely converge at the focal point regardless of where the light ray strikes the lens. In such a system, the focal point is truly a point in the mathematical sense and is a circle whose diameter is exactly zero. However, because light has a wavelength greater than zero, the light rays do not converge to a precise mathematical focal point, but instead converge within a small circle whose center is the focal point. The size of this small circle depends upon the diameter of the lens and the wavelength of light.

The small circle centered around the focal point is called the Airy disc in honor of Sir George Airy who first calculated its size. For a light ray of wavelength λ, a lens with diameter D and a focal lenth of f, the radius of the Airy disc is given by the equation

$$r = 0.00122\lambda f/D. \qquad (3.2.1)$$

The resulting radius is expressed in microns, λ is expressed in nanometers (nm), and the diameter D is in millimeters. The relationship between microns and millimeters is given by the equation

$$1\text{ micron} = 0.0001\text{ mm}. \qquad (3.2.2)$$

(Equation 3.2.1 may seem like it is mixing a lot of unlike units, but the scaling that is used here is more convenient than converting all length measurements to the same units. Thus, wavelengths are entered in nanometers, such as 550nm, rather than a more unwieldly measurement unit, such as 0.55 microns.)

Notice that equation 3.2.1 contains f/D, the focal ratio for the lens. Expressing equation 3.2.1 in arcseconds instead of microns, we have

$$r = \frac{0.2516433\lambda}{D}. \qquad (3.2.3)$$

Because light rays behave like waves, there is interference between light rays. The impact of this is that light intensity is not uniform around the focal point. Instead, there are concentric circles of light around the focal point with decreasing intensity away from the focal point. These concentric circles are called the Airy pattern. Figure 3.7 shows the Airy pattern. The dark spaces between the light circles represent areas where the light rays "interfere" with each other and cancel out each other. When there is no obstruction between the lens and the object being viewed, the Airy disc contains 84% of the light in the image. The next innermost ring contains 7.2% of the image's light.

The Airy disc has two important implications. First, since the Airy disc depends upon the wavelength of the light rays, it affects the emulsions which may be used on photograph plates for astrophotography. Secondly, the size of the Airy disc for a telescope sets a theoretical

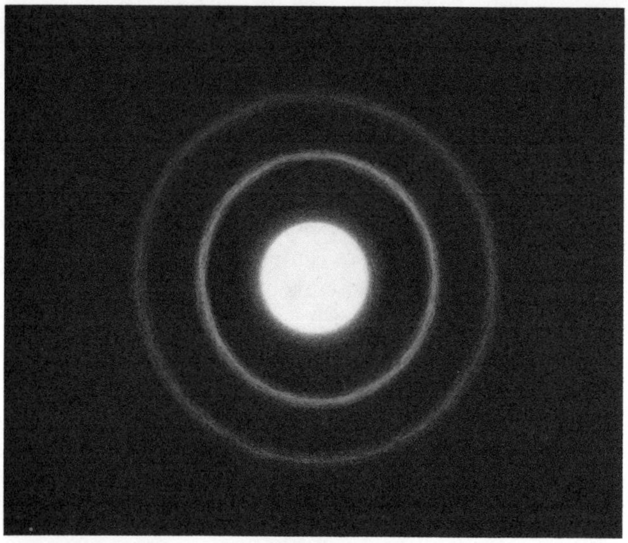

Figure 3.7: *The Airy pattern results from an uneven distribution of light intensity around the focal point. The Airy disc is the brightest circle centered on the focal point.*

limit on the telescope's resolving power. The resolution of a telescope is a measure of how well the telescope can distinguish between two stars that are very close together.

If two stars are so close together that their images fall within the Airy disc, they will appear as a single star when viewed through the telescope. The smaller the Airy disc, the closer two stars may be and still be distinguishable and hence the better the resolution of the telescope. From equation 3.2.1 and for a specific wavelength of light, the wider the aperture, the smaller the Airy disc. This is why larger aperture telescopes *theoretically* offer better resolution.

There are several ways to measure the resolution of a telescope depending upon exactly how resolution is defined. One definition is to multiply equation 3.2.3, the radius of the Airy disc itself, by 2.0 which says that a telescope can distinguish between two stars if they are separated by their own angular distance. Now equation 3.2.3 is dependent upon the wavelength λ. The human eye is most sensitive to light with a wavelength of 550nm (yellow light). Substituting this wavelength in equation 3.2.3 and multiplying by two gives

$$R_{lower} = \frac{276.8}{D} \qquad (3.2.4)$$

as one way to measure a telescope's resolving power. D is measured in millimeters in this equation and the result R_{lower} is in arcseconds. This equation represents a theoretical lower limit on resolution, for yellow light.

Lord Rayleigh noted that in actual practice, two stars can still be resolved when their angular distance

Aperture (mm)	Lower	Rayleigh	Dawes
25	11.07	5.54	4.63
50	5.54	2.77	2.32
75	3.69	1.85	1.54
100	2.77	1.38	1.12
125	2.21	1.11	0.93
150	1.85	0.92	0.77
200	1.38	0.69	0.58
250	1.11	0.55	0.46

Table 3.2: Resolving Power. *There are several different ways to measure and define a telescope's resolution. Resolution in this table is expressed in arcseconds.*

is roughly half the Airy disc. Again assuming a wavelength of 550nm, we have

$$R_{Rayleigh} = \frac{138.4}{D}. \qquad (3.2.5)$$

Another empirical measurement, the Dawes criteria, is

$$R_{Dawes} = \frac{115.82}{D}, \qquad (3.2.6)$$

but this equation is strictly only valid for white double stars when viewed through a 150mm telescope. Table 3.2 shows the resolutions for telescopes with several different size apertures.

All three equations given here express the same central idea. If the angular separation between two objects is less than R seconds of arc, the two objects will appear to be one in the image. If the angular separation is greater than R seconds of arc, it will be apparent in the image that two distinct objects are being viewed. When studying distant binary stars, the resolution is an important factor.

There are three points that must be kept in mind when discussing a telescope's resolving power. First, there are several ways to define resolution and it is important to know which definition is being used. Secondly, the equations presented above are defined only for a specific wavelength of light, namely 550nm light. Thirdly, resolution is not the same thing as contrast. Resolution measures the ability to distinguish between two different closely spaced objects while contrast measures the ability to distinguish between features in an image. Contrast is important for studying planetary details.

Suppose two objects have light intensities I_1 and I_2 with $I_1 > I_2$. If the light intensity varies between 0 and 1, then the mathematical definition of contrast is

$$C = 100\frac{I_1 - I_2}{I_1 + I_2} \qquad (3.2.7)$$

where C is expressed as a percentage. For example, if the

intensity of the first object is $I_1 = 1.0$ and the second is $I_2 = 0.0$, then the contrast C is 100%. If the second intensity is $I_2 = 0.25$, then the contrast is $C = 60\%$. oor contrast makes it difficult to distinguish planetary features even if the resolution is good. Further discussion of contrast is beyond the scope of this book. The interested reader is refered to the works cited in Appendix C.

From the equations presented so far, it would appear that an ideal telescope would have an infinitely large objective diameter, an infinitely long objective focal length, and an infinitely short eyepiece focal length. Obviously such a telescope is impossible to build. Once built, it is not possible to change the telescope's aperture size nor the characteristics of its objective lens. Only the eyepiece can be changed. But for a given aperture size and objective focal length, there are bounds on the useful magnification that can be obtained by changing the focal length of the eyepiece. This is true because of the finite size of the human eye and because a finite amount of light enters the telescope. As a general guideline, the maximum and minimum useful magnification are given by the equations

$$Max\ Power \approx 2.5D \tag{3.2.8}$$

and

$$Min\ Power \approx 0.13D. \tag{3.2.9}$$

In both equations, the objective diameter D must be expressed in millimeters. Once the maximum and minimum useful magnifications are known, the required focal length of the eyepiece can be determined from equation 3.1.6. As in the case of resolution, there are both theoretical and empirical equations that may be found to define limiting power. Other equations, in addition to the two given here, may be found in J. B. Sidgwick's *Amateur Astronomer's Handbook.*

As an example, suppose we have a 6-inch telescope whose objective lens has a focal length of 50mm. Calculate the maximum and minimum useful magnification and the required eyepieces.

1. Using the crossmultiplication techniques from the previous chapter, convert the aperture size, 6 inches, into millimeters. Since

$$1 \text{ inch} = 25.4 \text{ mm},$$

we have

$$6 \text{ inches} = 152.4\text{mm}.$$

2. Applying equation 3.2.8 gives

$$Max\ Power \approx (2.5)(152.4\text{mm}) = 381.$$

3. Applying equation 3.2. gives

$$Min\ Power \approx (0.13)(152.4\text{mm}) = 1\ .812.$$

4. Rounding the calculated magnifications gives a maximum useful magnification of 380 power and a minimum useful magnification of 20 power.
5. Applying equation 3.1.6, to obtain 380 power magnification requires a 3mm eyepiece.

$$f_{eye} = \frac{50\text{mm}}{380} = 2.5\text{mm}$$

6. Again applying equation 3.1.6, a 48mm eyepiece is required for 20 power magnification.

$$f_{eye} = \frac{50\text{mm}}{20} = 47.5\text{mm}.$$

Increasing the size of a telescope's aperture increases the amount of light that reaches the eyepiece. This makes stars appear brighter and is why more stars can be seen through a telescope than with the naked eye. Stars are classified according to their magnitude which is a measure of how bright they appear. The higher the magnitude number, the fainter the star. Since brightness is nonlinear, the magnitude numbers are also nonlinear. For example, a 1st magnitude star is 4000 times brighter than a 10th magnitude star instead of the expected 10 times brighter. With only the naked eye humans can see stars down to about magnitude 6.5.

Given an objective diameter expressed in millimeters, the magnitude of the faintest visible object is given by

$$m \approx 7.7 + 5 * \log_{10}(0.1D). \tag{3.2.10}$$

This equation calculates a telescope's limiting magnitude and very clearly points out that increasing the aperture size increases the amount of light entering the telescope and hence allows fainter objects to be seen. For example, a telescope with a 500mm aperture has a limiting magnitude of approximately 16.2 while a telescope with a 100mm aperture has a limiting magnitude of approximately 12.7.

All things considered, what is the optimum magnifying power to use for a given telescope? There is of course no hard and fast answer to this question because it depends greatly upon what is being viewed. However, a rough equation for optimum magnification, taking into consideration power, resolution, and contrast, is

$$Power_{opt} = 1.25D. \tag{3.2.11}$$

This equation assumes fairly low contrast (10%) and ideal viewing conditions. Notice that the theoretical optimum power lies roughly between the theoretical minimum and maximum useful power.

For an 8-inch (203.2mm) telescope, the optimum magnification power according to this equation is

$$Power_{opt} = (1.25)(203.2) = 254.$$

This may seem disappointingly low if one is used to thinking in terms of using the largest possible magnifying power!

The equations in this section provide a quantitative means for measuring a telescope's quality. It can be seen from the equations that the objective lens characteristics are of paramount importance. Thus, the remark in the introduction to this chapter that aperture size is generally more important than magnification power is borne out by the equations.

3.3 Image Aberrations

The previous section described physical limitations that apply equally to lenses and mirrors. However, there are other image aberration effects that are important considerations as well. Spherical aberration and chromatic aberration will be described briefly in this section. The interested reader is refered to works cited in Appendix C for a more rigorous discussion of these and other types of image aberrations.

Spherical aberration is an image distortion that occurs because not all light rays passing through a spherical converging lens (or reflected by a spherical concave mirror) pass exactly through the focal point. Some light rays pass in front of the focal point while other pass behind the focal point. Spherical aberration makes it impossible to precisely focus an image, but unless it is severe enough, the human eye may not be able to perceive it. But even though the human eye may not be sensitive enough to detect the presence of spherical aberration, it can be a serious limitation in astrophotography.

If the spherical aberration is smaller than the size of the Airy disc, no correction is needed because no amount of correction will improve the quality of the resulting image. Spherical aberration can be minimized through a combination of lenses which "over compensate" and "under compensate" to cancel out the spherical aberration. Spherical aberration in a mirror can be corrected by making the mirror parabolic in shape rather than spherical, but this can introduce other undesirable effects. Some reflectors, such as the Schmidt-Cassegrain, use a combination of a spherical mirror and a lens to correct for spherical aberration.

Unlike spherical aberration, chromatic aberration, exists only in refractive systems. (Note that an eyepiece, even for a reflector telescope, is a refractive system and is thus susceptible to chromatic abberation.) Chromatic aberration is an image distortion which occurs because light rays of different wavelengths are affected differently in the telescope system. The amount of distortion depends upon the wavelength being viewed and the refractive index of the lens. Since the wavelength of a light ray determines the color perceived by the eye, the effect is that stars appear as colored blurs.

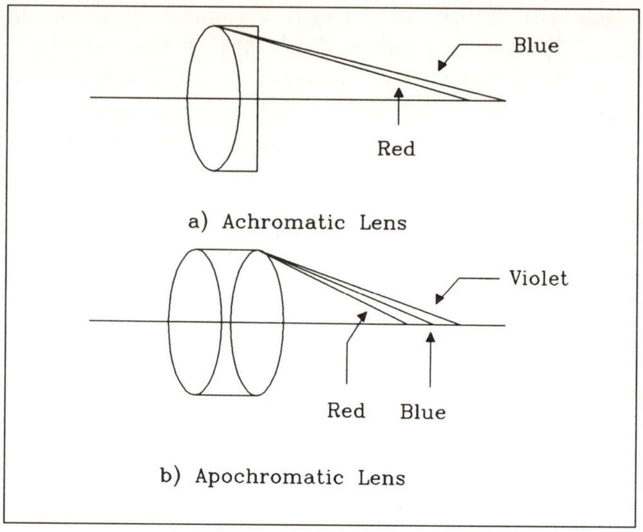

Figure 3.8: *Chromatic aberration can be minimized by using multiple lens configurations. This illustration shows a typical cemented achromatic lens configuration and a typical cemented apochromatic configuration.*

The easiest way to understand chromatic aberration is to realize that a lens behaves much like a prism. Light rays passing through a prism are refracted by different amounts depending upon their wavelength. This is why a white ray of light, which consists of multiple wavelengths, is decomposed into a spectrum of colored light rays. This property of light is not all bad because it is the fundamental principle which allows a spectrometer to work for determining the composition of stars.

There are two types of chromatic aberration. The first, longitudinal chromatic aberration, occurs when the "prism" effect is noticeable around the lens axis, that is, a line which passes through the center of the lens. Lateral chromatic aberration, the second type of chromatic aberration, occurs when the "prism" effect is noticeable around the edges of the lens. This effect can be easily seen as color fringes around an image viewed through a pair of cheap binoculars.

One way to reduce chromatic aberration is to use lenses with long focal lengths relative to the aperture size. This approach is limited to 8-inch or less aperture sizes because the focal length needs to be about 15 times the aperture size to sufficiently reduce chromatic aberration. For an 8-inch telescope, this means the main telescope body must be 10 feet long, 12.5 feet for a 10-inch telescope, and 25 feet for a 20-inch telescope!

Chromatic aberration can be reduced, but not completely eliminated, by using multiple lenses. Figure 3.8 shows a two lens and a three lens solution. The two lens design, called an achromatic lens, contains both a

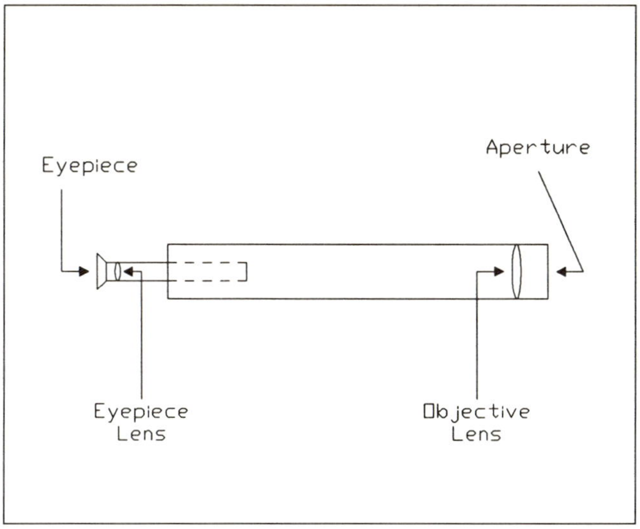

Figure 3.9: *The key components of a refractor telescope.*

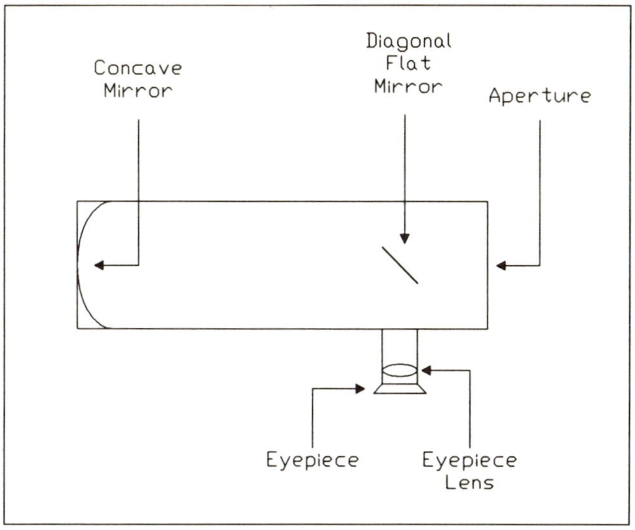

Figure 3.10: *The key components of a Newtonian reflector telescope.*

converging and a diverging lens. This lens is usually designed to reduce chromatic aberration by causing two wavelengths of light, usually red and blue, to converge as closely as possible to the focal point for yellow light. The three lens design, called an apochromatic lens, consists of two converging lenses and one diverging lens. This design is constructed so that three wavelengths of light (usually red, blue, and violet) converge as closely as possible to the focal point for yellow light.

The simplest approach to eliminating chromatic aberration is to use colored filters. This is the prefered approach for astrophotography.

3.4 Telescope Construction

There are two basic types of telescopes: refractors and reflectors. Figure 3. illustrates a typical refractor telescope. A relatively large glass lens, called the objective lens, is placed at one end of a long tube. At the other end, a small movable tube contains an eyepiece lens. The eyepiece can be moved in or out of the main telescope tube to focus the image. The eyepiece lens is normally removable so that different eyepieces can be inserted to change the telescope's magnification.

There are two basic types of refractors that the amateur is likely to encounter. The doublet refractor contains an achromatic objective lens to reduce the effects of lateral chromatic aberration. A doublet refractor still suffers from longitudinal chromatic aberration. The triplet refractor contains an apochromatic objective lens to further reduce lateral chromatic aberration and to reduce longitudinal chromatic aberration. However, the triplet refractor is much more expensive to build and probably can not be justified unless the telescope is to be used for astrophotography.

The best material to use for a doublet's objective lens is a manmade optical material called fluorite. As shown in Table 3.1, fluorite has a low refractive index, lower than the crown and flint glass used in many lens systems. Fluorite offers substantially better color correction characteristics, but is much more expensive, is limited in the practical size of the aperture, and is highly prone to adverse weathering effects.

A typical reflector telescope is shown in Figure 3.10. This design is called a Newtonian reflector. Light entering the aperture strikes a concave mirror which magnifies the image and reflects the light onto a flat mirror. The flat mirror in turn reflects the light into a movable eyepiece which focuses the image. Notice that the reflector is open at the aperture end which allows dust and air to enter the telescope body. A refractor on the other hand is sealed at both ends.

The Newtonian reflector has long been a popular telescope for amateurs because it is relatively easy for an amateur to build. It is also less expensive than an equivalent refractor telescope, is more compact, and has no chromatic aberration. However, the Newtonian is limited to relatively short focal ratios and is susceptible to adverse atmospheric viewing conditions due to the open telescope body.

The Cassegrain reflector, shown in Figure 3.11, is a commonly available design that is much more compact than a refractor or Newtonian reflector. Light entering the aperture strikes the primary objective mirror and is reflected onto a much smaller secondary convex mirror. From the secondary mirror, the light is reflected through a hole in the primary mirror into the eyepiece. The major disadvantage with a Cassegrain reflector is that

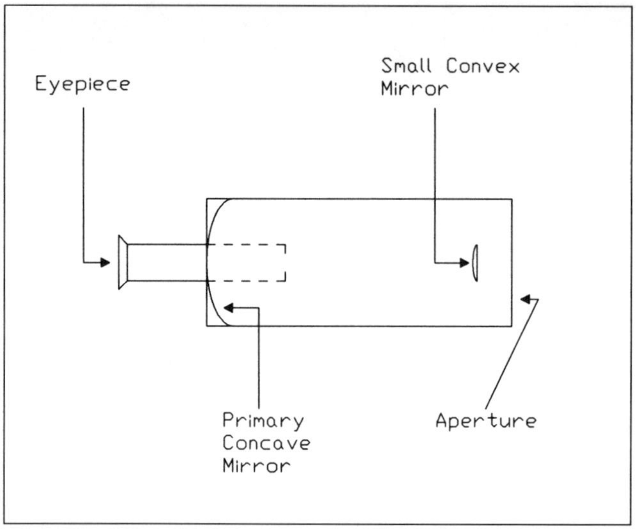

Figure 3.11: *The key components of a Cassegrain telescope. Popular today are modified Cassegrains called catadioptrics that have thin correcting lenses at the tube's top (right in the diagram).*

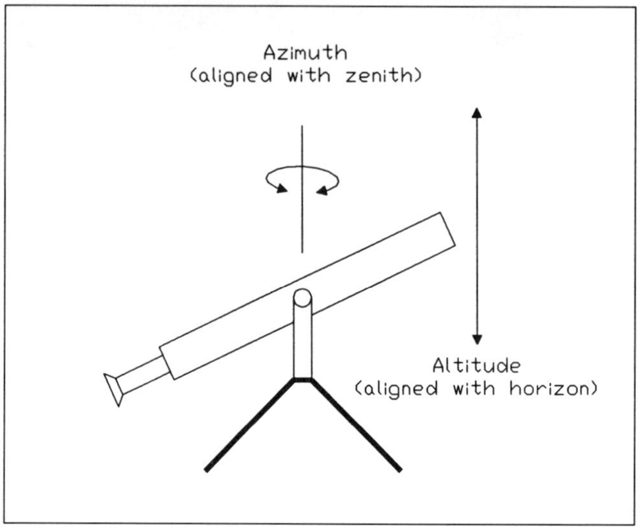

Figure 3.12: *Alignment of an altazimuth mounting is with respect to an observer's horizon coordinate system. This type of mount does not lend itself to tracking the stars because its axes cannot be made to coincide with the Earth's.*

the secondary mirror blocks some of the light entering the telescope. This decreases the brightness of the resulting image and decreases image contrast and sharpness. The Cassegrain has the advantage of having a relatively long equivalent focal length due to the addition of the secondary convex mirror.

Two variations on the basic Cassegrain design are the Schmidt-Cassegrain and Maksutov-Cassegrain designs. In a Schmidt-Cassegrain telescope, a Schmidt corrector lens is placed at the open end of the telescope tube to correct for the spherical aberration of the primary mirror. The lens introduces chromatic aberration, but it usually is not severe. An advantage of this variation is that it has a closed tube body and is compact, although not as compact as the Casegrain.

The Maksutov-Cassegrain design differs from the Schmidt-Cassegrain in that the Schmidt corrector lens is replaced by a meniscus corrector, a strongly bent lens. The meniscus lens corrects the problem of spherical aberration by introducing a "negative" aberration to counteract that of the primary mirror.

Commercially available telescopes offer a wide range of variations on the two basic telescope types and the design variations described. Each variation represents an attempt to improve some characteristic of the telescope, usually at the expense of another characteristic. Regardless of which type of telescope is chosen, the fundamental limitations described in preceding sections should be kept in mind.

3.5 Telescope Mountings

Even when low magnifications are involved, the slightest movement of a telescope can cause the image to appear to "vibrate" in the eyepiece. This vibration makes viewing difficult at best and can be very frustrating. For this reason the telescope should be mounted on a solid tripod which can hold the telescope motionless. In practice there is almost always some backlash in gearing which does cause some "jittering" until the motion settles down.

Besides providing a stable platform for the telescope, mountings can aid in positioning the telescope for viewing an object. Commercial mountings are generally either altazimuth or equatorial. Both of these names correspond to coordinate systems that are used to determine the location of a star as will be discussed in Chapter 5.

The altazimuth mounting allows a telescope to move on two axes. The first axis allows the telescope to move in a circle which is parallel to the horizon. The second axis allows the telescope to move in a circle which is perpendicular to the horizon. Motion in these two planes allows the telescope to point in any direction. Figure 3.12 is a schematic diagram of the altazimuth mounting.

Equatorial mountings also allow the telescope to move in axes which are perpendicular to each other. Instead of being related to the observer's horizon, one

Figure 3.13: *This altazimuth mount is in the form of the popular and easy-to-build Dobsonian design.*

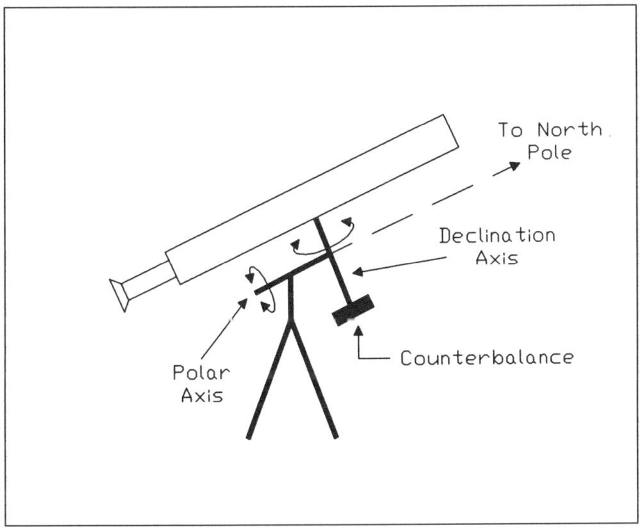

Figure 3.14: *Alignment of an equatorial mount is with respect to the Earth's axis of rotation. Note than when aligned with the Earth's axes, a counter rotating motor attached to the polar axis will cause objects in the field of view to appear stationary.*

Figure 3.15: *An equatorially mounted refractor telescope. Ragland Observatory, Richmond Astronomical Society by Randy Tatum.*

equatorial axis is parallel to the Earth's axis of rotation while the second axis is perpendicular to the Earth's axis of rotation. This type of mount is illustrated in Figure 3.14.

The altazimuth is the easiest and cheapest mounting to build. This mounting is also very simple to learn to use. However, the altazimuth mounting requires a constant readjustment of both axes to track an object as it moves across the sky. While the equatorial mounting is more expensive to build, tracking an object across the sky can be accomplished by adjusting only one axis of rotation instead of two. Hence, with an equatorial mount a motor can be attached to one axis to allow for automatic tracking of an object. Most people prefer the equatorial mounting for this reason. In fact, astrophotography is done almost exclusively with equatorial mounts.

An equatorial mount must be aligned with the Earth's axis of rotation, with its polar axis pointing towards the North Celestial Pole. If the telescope is to be permanently installed at one location, the alignment of an equatorial mount is not a major undertaking. However, if the telescope site varies nightly, precise alignment (in the dark!) can be frustrating for the inexperienced.

The operation of both the equatorial and altazimuth mountings will make more sense when coordinate systems are described in Chapter 5. For now it suffices to be aware that different mountings exist which have advantages and disadvantages.

3.6 Eyepieces

The telescope eyepiece is just as important a consideration as the selection of a mounting or the telescope itself. The modern astronomer has a wide variety of eyepieces, also called oculars, to choose from. Eyepieces are multilens designs because a single lens design suffers greatly from chromatic aberration. The Huygens eyepiece, originally designed in 1703, was the first two lens design. Unfortunately, the simple Huygens design suffers greatly from spherical aberration. The Ramsden lens reduces the spherical aberration of the Huygens eyepiece by a factor of 8 but suffers from lateral chromatic aberration. More recent designs, such as the Kellner, Plössl, Abbe, König, and Erfle, greatly reduce both spherical and chromatic aberrations. The Abbe lens is also called an orthoscopic lens and offers exceptionally low distortion. The Erfle design contains 5 lenses and offers a wide field of view. Table 3.3 gives typical apparent field sizes for several eyepiece designs.

Design	Apparent Field
Abbe	50°
Erfle	70°
Hastings	25°
Huygens	50°
Kellner	50°
König	50°
Plössl	50°
Ramsden	40°
Steinheil	25°
Tolles	12°

Table 3.3: Eyepiece Designs. *Modern astronomers can select from a wide variety of eyepiece designs with different optical characteristics.*

In addition to affecting the telescope's magnifying power, as shown in equations 3.1.5 and 3.1.6, the eyepiece affects the telescope's field of view. The true field of view, expressed in degrees, is a measure of how much of the sky is visible through the eyepiece. The apparent field of view is a fixed characteristic of the eyepiece selected. The wider the field of view, the greater the amount of sky that can be seen without repositioning the telescope.

Power	Field of View
10	4°30′00″
20	2°15′00″
25	1°48′00″
50	0°54′00″
75	0°36′00″
100	0°27′00″
150	0°18′00″
200	0°13′30″
250	0°10′48″
500	0°05′24″

Table 3.4: True Field of View for a Set of 45° Eyepieces. *Increasing magnification decreases the field of view while decreasing magnification increases the field of view.*

The true field of view is determined by two factors: the magnifying power used and the eyepiece used. For an eyepiece with an apparent angular field of E degrees and power of M, Sidgwick gives the following equation for calculating the true field of view θ:

$$\theta = \frac{E}{M}. \tag{3.6.1}$$

Assuming an average value of 45° for an eyepiece's apparent angular field, Table 3.4 gives the resulting true field of view for several magnification powers. Notice that indeed the field of view decreases as magnifying power increases.

As equation 3.6.1 demonstrates, magnification has an adverse affect on true field of view. As magnifying power is *increased*, the resulting true field of view is *decreased* making it very much dependent upon the eyepiece used. Decreased field of view is another reason why magnifying power should not be the number one consideration in selecting and using a telescope. As a general rule, one should use the smallest possible power to suit the viewing task at hand.

3.7 Selecting the "Best" Telescope

Both the refractor telescope and the reflector telescope have distinct advantages and disadvantages. Reflectors, for large apertures, are generally less expensive and easier to build than refractors. On the other hand, refractors are less prone to atmospheric disturbances (temperature, air turbulence) and are easier to maintain than reflectors since difficult to reach optical surfaces require reflectors to be dismantled for cleaning. Refractors are also easier to maintain than reflectors because reflector mirrors may need to be recoated after a number of years.

Refractor telescopes suffer from chromatic aberration while reflectors do not. The undesirable obstruction of a secondary mirror is the price paid for selecting an objective mirror over an objective lens. Because of this obstruction, the resolving power of a refractor is generally better than that of a reflector. Reflectors are more compact which may be an important consideration for the amateur.

What constitutes the "best" telescope for the amateur? Texereau, in his book *How to Make a Telescope*, suggests the following characteristics for a home made telescope:

1. It should be a reflector because it is easier to construct than a refractor and is more compact than an equivalent refractor.
2. It should be a Newtonian reflector because it is easier for the beginner to construct. (This suggestion may not be appropriate if a telescope is purchased instead of built from scratch.)
3. The aperture should be an 8-inch aperture. This represents a reasonable overall compromise.
4. The focal ratio should be between $f/6$ and $f/8$ to decrease the size required for the telescope body.
5. An altazimuth mounting should be used since it is easier to set up and is low in cost. Dobsonian mounts are especially attractive for the beginning amateur astronomer.

Once the "best" telescope is selected, it still may not live up to expectations. The reason is that the Earth's atmosphere introduces various difficulties for all telescopes that make viewing less than optimum. A telescope will usually perform at its peak for only fleeting moments and in fact, the atmosphere can decrease viewing performance by a factor of 10 over defects in the optical process of creating the telescope in the first place! The larger the telescope aperture, the more pronounced are the effects. Atmospheric turbulence causes a twinkling effect when viewing stars, causes a loss of resolving power, and reduces contrast.

Some adverse viewing conditions occur because of the telescope itself. Tube currents inside the body of an open telescope, as in the Newtonian reflector, cause images to appear fuzzy and even to oscillate. The problem occurs whenever the aperture is more than about 30cm. When the temperature of the telescope body or lens differs from that of the atmosphere, dewing presents a potential problem. Drops of dew essentially introduce another optical surface and can reduce light grasp as well as introduce image aberrations described above.

Perhaps for many viewers, the presence of light from street lights or the city are an even more serious factor. This factor makes portability of the telescope even more important! The presence of extraneous light lessens the intensity and contrast of images seen through the telescope. Smog or other atmospheric pollutants have the same effect by decreasing all light entering the telescope, including the desired light from stars or planets.

3.8 About the Program

TELSCOPE is provided to illustrate the computations described in this chapter. TELSCOPE requires entry of lens sizes which are sometimes easier to express in inches and sometimes easier to express in millimeters. When appropriate, the program will automatically convert from inches to millimeters but you must indicate to the program which units you are using by adding a suffix of `I` to indicate inches or `M` to indicate metric units. (TELSCOPE is not case sensitive so lowercase letters will work as well.) Thus, an objective lens for a 6-inch telescope would be entered as

```
6I
```

while an eyepiece with a 25mm lens would be entered as

```
25M.
```

Unless the program specifies otherwise, you may freely mix metric and English units.

In addition to performing the basic calculations described in this chapter, the program will also compute a data table which summarizes the characteristics of a given telescope. The data table produced summarizes magnifying power for a given objective lens and up to 10 eyepieces. The theoretical limits (resolution, star magnitude, and magnification) are computed as well as the power and field of view for several different eyepieces.

It is important to note that most versions of `BASIC` do not include a base 10 logarithm function. TELSCOPE takes care of this deficiency by defining a function, `FNLN`, which computes base 10 logarithms. `FNLN` is defined by

$$\mathtt{FNLN}(x) = \frac{\log x}{\log 10}$$

where log is the Napierian or natural (base e) logarithm.

3.9 Sample Problems

1. Given a 4-inch telescope with a 750mm focal length, calculate the focal ratio.
(Ans: $f/7.4$)
2. Calculate the focal length of a $f/4.5$ 10-inch telescope.
(Ans: 1143 mm)
3. Calculate the aperture size of a $f/6$ telescope whose objective lens has a focal length of 1,000mm.
(Ans: 167mm (6.5-inch))
4. Calculate the magnification power achieved using a 100mm objective lens and a 6mm eyepiece.
(Ans: 17)
5. Calculate the magnification power achieved using a 200mm objective lens and a 28mm eyepiece.
(Ans: 7)

6. Calculate the eyepiece required to achieve 100 power when the objective lens has a focal length of 100mm.
(Ans: 1mm)
7. Calculate the eyepiece required to achieve 25 power when the objective lens has a focal length of 200mm.
(Ans: 8mm)
8. Calculate the Airy disc radius for an 8-inch telescope whose focal length is 100mm. Assume the wavelength of light is 550nm.
(Ans: 0.33 microns)
9. Calculate the Airy disc radius for a $f/10$ telescope for 550nm light.
(Ans: 6.7 microns)
10. Calculate the resolution, in arcseconds, for a 4-inch telescope.
(Ans: $R_{lower} = 2.72$, $R_{Rayleigh} = 1.36$, $R_{Dawes} = 1.14$)
11. Calculate the resolution, in arcseconds, for an 8-inch telescope.
(Ans: $R_{lower} = 1.36$, $R_{Rayleigh} = 0.68$, $R_{Dawes} = 0.57$)
12. Calculate the theoretical minimum, optimum, and maximum useful power for a 4-inch telescope.
(Ans: Min=13, Opt=127, Max=254)
13. Calculate the theoretical minimum, optimum, and maximum useful power for a 6-inch telescope.
(Ans: Min=20, Opt=191, Max=381)
14. Calculate the limiting magnitude for a 6-inch telescope.
(Ans: 13.6)
15. Calculate the limiting magnitude for a 4-inch telescope.
(Ans: 12.7)
16. Calculate the field of view for a typical Plössl eyepiece at 100 power.
(Ans: 0.5°)
17. Calculate the field of view for a typicl Erfle eyepiece at 15 power.
(Ans: 4.7°)

Chapter 4

Time Conversions

Everyone has a concept of what time is. Yet explaining what this abstract concept means is like trying to explain the color red to a blind person. Referring to a dictionary for help will only compound the problem! Time, whether we can adequately describe it or not, plays an important role in our daily lives. Time is also an important quantity in astronomy as well as in the study of science in general.

The present chapter describes how to perform some important time conversions. Such conversions are required for determining star locations and the location of objects within the solar system. This will be more apparent later but the basic idea is that knowing where an object was at some instant and how much time has elapsed since then, the object's new position can be calculated. The calculations required involve converting dates into a more convenient form and converting between local time and "star" time.

The subject of time can be very confusing. The discussion below will describe 5 ways to define a year, 4 ways to define a month, and over 6 ways to refer to the time of day! In addition, just to make things interesting, there are two different calendars to worry about! For the moment concentrate on how to perform the conversions rather than why they are needed.

Figure 4.1: *This sundial is on the North wall of the Club House at Stellafane. The shadow falls on the 11 hour line and since this picture was taken in early August one finds from the correction table running down the right side of the dial that it is necessary to add 6 minutes to get the correct time.*

4.1 Defining Day, Month, and Year

Among the smaller units of time that can be gauged by astronomical events is the day. We normally consider a day as beginning and ending at midnight but this is really an arbitrary decision. Another choice might be to count a day as beginning and ending with sunset. The Jewish nation of Biblical times counted days in this manner. In any event, a day is normally defined as the time it takes for the Sun to return to the same location in the sky that it was the day before.

A day as described in the previous paragraph is called a solar day. Time measured by the solar day is called apparent or solar time. A sundial indicates solar time. Unfortunately, a solar day is not uniform in length since the Sun's motion is non-uniform. Two factors account for this irregularity in motion. First, the Earth moves about the Sun in an elliptical orbit instead of a circular orbit. Thus, as explained by Kepler's laws, the Earth moves faster as it get closer to the Sun and slower as it gets farther away from the Sun. This causes the rate of the Sun's apparent motion along its ecliptic path in the sky to vary during the year. Secondly, the Earth's axis is tilted at an angle to the plane of its orbit around the Sun. This has the main effect of making the amount of daylight vary with the seasons.

To avoid the difficulties caused by irregularities in the Sun's apparent motion, astronomers define a fictitious Sun, called the mean Sun, which moves at a uniform rate along the Earth's equator. A day defined by the motion of this fictitious Sun is called a mean solar day. A mean solar day is exactly 24 hours in length while as indicated before, a solar day can vary by several minutes.

Formally speaking, a solar day is defined to be the time interval between two successive transits of the Sun across the observer's meridian. The concept of a meridian and an observer's meridian will be explained more fully in Chapter 5. For now think of the observer's meridian as a semicircle which passes through three points: the North Pole, the South Pole, and the point that is directly overhead the observer. Note that the observer's meridian depends upon where the observer is located on the Earth. In this context, a transit means to cross the observer's meridian. Roughly speaking then, a solar day begins when the Sun is at its highest point in the sky (about noontime) and ends the next day when it returns to its highest point. In a similar fashion, a mean solar day is defined to be the time between two successive transits of the mean Sun across the observer's meridian.

One advantage of the mean solar day is that the mean Sun returns to the same point in the sky in exactly 24 hours. However, if one is interested in observing the stars, a star does not return to the same position in the sky in 24 hours as measured by a mean solar day clock. This is true because at the end of a mean solar day, the Earth has advanced in its orbit around the Sun by about 1°. The effect is as if the stars move with respect to the Sun. This difficulty can be avoided if a day is measured with respect to the stars instead of the Sun. This is in fact what astronomers have done. A sidereal day is defined to be the time interval between two successive transits of a fixed star across the observer's meridian. Time measured according to a sidereal day is sometimes loosely called "star" time. At the end of exactly 24 sidereal hours, 1 sidereal day, a star returns to the same position that it was one sidereal day earlier.

Clocks can be built which measure time according to a solar day, a mean solar day, or a sidereal day. Sundials measure time based upon an apparent solar day. Our modern clocks measure time based upon a mean solar day. Special clocks, called sidereal clocks, are used for astronomical purposes and are regulated by the motion of the stars. A sidereal day is slightly shorter than a mean solar day. 23^h56^m of mean solar time is equal to about 24 hours of mean sidereal time. Henceforth unless otherwise noted, a 24-hour day will refer to a mean solar day.

Just as the apparent motion of the Sun can be used to define time intervals, the Moon's orbit is used as the basis for defining the month. And just as there are several ways to define a day, there are several ways to define a month. On a simple level, a month is the time it takes for the Moon to complete one orbit around the Earth. At a similar simple level, a day is the time it takes for the Sun to complete one apparent trip around the Earth, and a year (described below) is the time it takes the Earth to complete one orbit around the Sun. (Actually, a day is the time for the Earth to rotate once on it's axis but this is conceptually equivalent to the Sun revolving about the Earth.)

To measure a month, it is necessary for the Moon to complete one orbit with respect to some reference point. If the reference point is the stars, it takes the Moon 27.3217 days to complete one orbit around the Earth and return to the same position against the background of stars. This month is called a sidereal month which represents a complete trip of the Moon around the Earth. When the Sun is used as the reference point, the Moon takes 29.5306 days to complete an orbit and defines the synodic month, which is the "phase" month: New Moon to New Moon, Full Moon to Full Moon, and so on. However, the lunar month is rarely used for civil purposes except for certain religious calendars, such as the Jewish and Moslem. In such calendar systems, based on lunar months, a given religious festival will always occur during the same lunar phase.

In defining a year, one can think of the Earth orbiting the Sun or the Sun completing an orbit around the Earth. The definition of the tropical year is easiest to visualize by assuming that the Sun orbits the Earth. The Sun's orbit crosses the plane of the Earth's equator twice a year. These two times are called the equinoxes and occur around March 21 (vernal equinox) and September 22 (autumnal equinox). On these two dates the length of the day and night are approximately equal. The tropical year is the time between two successive vernal equinoxes and is equal to about 365.2422 days. Unless otherwise noted, a year is assumed to mean a tropical year.

The tropical year suffers the disadvantage of having a fractional number of days in it. To circumvent this difficulty, the civil year is defined to be either 365 or 366 days in length depending upon whether or not it is a leap year. The Besselian year is a refined version of the tropical year and is used in astronomical calculations. When the stars are used as a reference point, the sidereal year is defined which is equal to 365.2564 days. Starting from the point that the Earth is closest to the Sun and back to it, the anomalistic year is defined and is 365.2596 days in length.

4.2 Calendar Dates

Because of the various possible definitions of a day, month, and year, it is small wonder that there are multiple ways to make up a calendar. Early calendars were based upon the apparent motion of the Sun and Moon and did not account for the fact that the Earth orbits the Sun in a fractional number of days. Thus, if a year is defined to be 365 days in length, the year is short by 0.2422 days. After 10 years, the calendar is wrong by 2 days. This was hardly noticeable to early man and would not be noticeable to the average person today. However, after 100 years, the calendar is wrong by 24 days and wrong by 242 days after 1,000 years. With this kind of error, it would soon be apparent that the seasons were occurring in the wrong months (winter in July, summer in December, etc.) when compared to recorded history. Such a calendar system is very inconvenient indeed.

The Roman emperor Julius Caesar introduced an improved calendar in about 46 B.C. His calendar, called the Julian calendar, assumes a year that is 365.25 days in length so that an additional day is added to the calendar every fourth year. So, any year that is evenly divisible by 4 contains an extra day and is called a leap year. This calendar is off by only 0.0078 days a year and is incorrect by about a week every 1,000 years. Close, but not close enough.

In 1582, Pope Gregory XIII was informed that the existing Julian calendar was incorrect by about 2 weeks. He established a new calendar system which retained the concept of a leap year, but modified the definition of a leap year. With this system, a century year (i.e., a year ending in 2 zeroes) is a leap year only if it is divisible by 400. This is the Gregorian calendar and is the calendar commonly in use today. The length of a year using the Gregorian calendar is 365.2425 days so that there is an error of only 0.0003 days a year. With the improvements made by Pope Gregory XIII, the calendar is only off by 3 days every 10,000 years. It is very unlikely that another calendar reform will occur for quite some time.

As an interesting sidelight, the Gregorian calendar was put into effect on October 15, 1582. At that time the 10 days between October 5 and October 14 were abolished. (In one sense, there never was an October 6, 1582!) The Gregorian calendar is in use by most countries of the world today, but England and the American colonies did not accept the new calendar until 1752. In that year, there were actually riots in England by protesters who demanded that they be given back their "missing" days.

There are several other calendars that have existed in the past or are in use even today. The Chinese calendar began in 2397 B.C. and is based upon 60 year cycles. The ecclesiastical calendar is used by some Roman Catholic and Protestant countries to reckon years beginning with Advent Sunday (the Sunday closest to the last Sunday in November). Calendars in use by the Hindus, Hebrews, and Moslems base years upon important religious events. Hindus divide the year into 12 months with the months based upon the signs of the zodiac. The Hebrew or Jewish calendar begins at 3761 B.C. with a cycle of 19 years in which the 3rd, 6th, 8th, 11th, 14th, 17th, and 19th years are leap years. The date 3761 B.C. was chosen as the date of the Creation. Moslem countries use the Moslem calendar which dates from July 16, 622 A.D. when the prophet Mohammed fled from Mecca to Medina. For the remainder of this book, however, only the Julian and Gregorian calendars will be of further interest.

Figure 4.2 *About 46 B.C. Julius Caesar introduced a calendar with a leap year.*

Determining whether or not a year is a leap year in the Gregorian calendar is a simple matter. To be a leap year, the year must satisfy two conditions.

1. The year must be evenly divisible by 4.
2. If the year is a century year, it must be evenly divisible by 400.

The year 1906 is not a leap year because it is not evenly divisible by 4. The year 1800 is evenly divisible by 4 but not by 400 so it is not a leap year either. The year 1600 is a leap year because it is a century year which is evenly divisible by 400.

4.3 Julian Day Numbers

In many astronomical calculations, it is necessary to know the number of elapsed days between two events. To facilitate calculating elapsed days, Julian day numbers are often used. When a date is converted to a Julian day number, the resulting number is frequently called the Julian date. (This is not the same as the Julian calendar date.) The Julian day number for a given date is the number of days which have elapsed since noon at Greenwich, England on January 1, 4713 B.C. It is important to realize that a Julian day number begins at $12^h0^m0^s$ (noon) Greenwich Mean Time (GMT). Converting time to GMT will be explained shortly.

Before describing the conversion process, one other detail needs to be explained. The Julian date also accounts for the time of day. To accomplish this, the time of day is expressed as a fractional part of the day and added to the day of the month. A couple of examples will illustrate the point.

Consider expressing 6:00 a.m., February 14 as a fractional day.

1. Express the time in decimal format using the techniques from Chapter 2. In this case, 6:00 a.m. becomes 6.0^h in decimal format.
2. Divide the decimal hours by 24. This will give the time as a fractional part of the day.

$$\frac{6.0}{24} = 0.25 \text{ days.}$$

3. Add the time as a fractional part of the day to the day so that we have February 14.25 as the date.

As a second example, consider expressing $14^h33^m36^s$ on March 21 as a fractional day.

1. In decimal format, $14^h33^m36^s = 14.56^h$.
2. Dividing by 24, we have

$$\frac{14.56}{24} = 0.606667 \text{ days.}$$

3. Adding this to the date gives March 21.606667.

Care must be taken in implementing a procedure which will produce the Julian date. The Julian dates can be rather large numbers and will cause roundoff problems unless double precision is used. For example, the Julian date corresponding to noon (GMT) on January 3, 1985 is 2,446,069.0. The programs for this chapter and succeeding chapters use double precision to minimize the effects of roundoff error.

Converting a calendar date to a Julian date can be done in 5 steps. Assume that the calendar date is $Month/Day/Year$. $Month$ is a number ranging from 1 to 12 with 1 = January, 2 = February, ..., 12 = December. Day is the day of the month and includes the time of day as a fractional part of the day. $Year$ is positive to indicate A.D. dates and negative to indicate B.C. dates. The necessary steps are listed below and an example problem is solved at the same time. In parentheses at the end of each step is the result that is obtained for the date January 3, 1985 at 0^h (midnight).

1. If $Month > 2$, set $Y = Year$ and $M = Month$. Otherwise, $Y = Year - 1$ and $M = Month + 12$. (Ans: $Y = 1984$, $M = 13$)
2. If $Year < 0$, set $T = 0.75$ else set $T = 0$. (Ans: $T = 0$)
3. Determine if the date is Gregorian or not. Dates before October 15, 1582 are not while all other dates are Gregorian. (Ans: 1/3/1985 is Gregorian)
4. If the date is Gregorian, compute

$$A = \texttt{FIX}\left(\frac{Y}{100}\right)$$

and

$$B = 2 - A + \texttt{FIX}\left(\frac{A}{4}\right).$$

If the date is not Gregorian, set $A = 0$ and $B = 0$. (Ans: $A = 19$, $B = -13$)

5. Compute

$$\begin{aligned} JD &= B + \texttt{FIX}(365.25 * Y - T) + \\ &\quad \texttt{FIX}[30.6001 * (M + 1)] + Day + \\ &\quad 1,720,994.5 \end{aligned}$$

to complete the calculation. (Ans: $JD = 2,446,068.5$)

Note that if the resulting Julian date is an integer, the time of day was noon (GMT) while if the Julian date ends in 0.5, the time of day was midnight.

The modified Julian date is sometimes used to avoid the difficulties of using the large numbers that are often produced from the above conversion process. The modified Julian date is the number of days since 0^h on November 17, 1858 and is equal to $JD - 2,400,000.5$. The reason for choosing 0^h November 17, 1858 should be obvious because of its Julian day number (2,400,000.5).

Converting a Julian day number back to the corresponding calendar date requires 10 steps. Assume JD is the Julian date. The Julian date 2,400,000.5 will be used to illustrate the process which follows.

1. Add 0.5 to the Julian date. That is, $JD = JD + 0.5$. (Ans: 2,400,001.0)

2. Compute $I = \texttt{FIX}(JD)$ and $F = \texttt{FRAC}(JD)$ where the FRAC function returns the fractional part of its argument. Thus, FRAC(1.9) is 0.9 and FRAC(−6.3) is 0.3 (not −0.3).
(Ans: $I = 2,400,001$, $F = 0.0$)
3. If $I > 2,299,160$, set
$$A = \texttt{FIX}\left(\frac{I - 1,867,216.25}{36,524.25}\right)$$
and set
$$B = I + 1 + A - \texttt{FIX}\left(\frac{A}{4}\right).$$
Otherwise set $B = I$.
(Ans: $A = 14$, $B = 2,400,013$)
4. Set $C = B + 1524$.
(Ans: $C = 2,401,537$)
5. Compute
$$D = \texttt{FIX}\left(\frac{C - 122.1}{365.25}\right).$$
(Ans: $D = 6574$)
6. Compute $E = \texttt{FIX}(365.25 * D)$.
(Ans: $E = 2,401,153$)
7. Compute
$$G = \texttt{FIX}\left(\frac{C - E}{30.6001}\right).$$
(Ans: $G = 12$)
8. The day is given by
$$Day = C - E + F - \texttt{FIX}(30.6001 * G).$$
(Ans: $Day = 17$, 0^{h})
9. If $G < 13.5$, $Month = G - 1$.
If $G > 13.5$, $Month = G - 13$.
(Ans: $Month = 11$)
10. If $Month < 2.5$, $Year = D - 4,715$.
Otherwise, $Year = D - 4,716$.
(Ans: $Year = 1858$)

If the year in step 10 is negative, the resulting date is B.C.

4.4 Other Date Calculations

Julian day numbers are convenient for calculations involving dates. To calculate the number of elapsed days between two dates, simply subtract the corresponding Julian dates. One possible use for this technique is to calculate how many days have elapsed since the beginning of the year. For a specific date, the elapsed days since the beginning of the year is called the day number or day of the year.

It is possible to compute the day number without using Julian dates. Two steps are required.

1. If $Year$ is a leap year, set $T = 1$.
Otherwise set $T = 2$.
2. The day number is given by the equation
$$\begin{aligned} N &= \texttt{FIX}\left(\frac{275 * Month}{9}\right) - \\ &\quad T * \texttt{FIX}\left(\frac{Month + 9}{12}\right) + \\ &\quad Day - 30. \end{aligned}$$

For example, the day number for March 9, 1981 is 68. This date is 68 days into the year.

Given the day number and the year, converting back to the calendar date is only a little more complex. Assume that the day number is N and for the purposes of this example, convert $N = 68$ for the year 1981 back to the calendar date. The algorithm is:

1. If $Year$ is a leap year, set $A = 1523$.
Otherwise, $A = 1889$.
(Ans: $A = 1889$)
2. Calculate
$$B = \texttt{FIX}\left(\frac{N + A - 122.1}{365.25}\right).$$
(Ans: $B = 5$)
3. $C = N + A - \texttt{FIX}(365.25 * B)$.
(Ans: $C = 131$)
4. Compute
$$E = \texttt{FIX}\left(\frac{C}{30.6001}\right).$$
(Ans: $E = 4$)
5. If $E < 13.5$, then $Month = E - 1$.
Otherwise, $Month = E - 13$.
(Ans: $Month = 3$)
6. $Day = C - \texttt{FIX}(30.6001 * E)$.
(Ans: $Day = 9$)

Sometimes it is interesting to know what day of the week a certain date falls on. For instance, it might be amusing to determine what day of the week that someone was born on. The Julian day number provides a simple way to calculate the day of the week. Using February 7, 1985 as an example, we have:

1. Calculate the Julian date, JD, for the given date at 0^h. (That is, make sure not to include the time of day as a fractional part of the day in the date. Thus $Day = 7$ and not 7.xxxx)
   (Ans: $JD = 2,446,103.5$)
2. Compute
   $$A = \frac{JD + 1.5}{7}.$$
   (Ans: $A = 349,443.571429$)
3. $B = 7 * \mathtt{FRAC}(A)$.
   (Ans: $B = 4.000003$)
4. $N = \mathtt{ROUND}(B)$ where the ROUND function rounds a number to the nearest integer.
   (Ans: $N = 4$)

The resulting number, N, from the above calculation gives the corresponding day of the week. $N = 0$ is Sunday, $N = 1$ is Monday, ..., $N = 6$ is Saturday.

4.5 Time of Day

There are several ways to refer to the time of day for two reasons. First, the time of day depends upon how the day is measured. Recall that a day can be a solar day, mean solar day, or sidereal day. Secondly, since the Earth rotates, the time of day depends upon where the observer is located.

As pointed out earlier, the length of a solar day depends upon seasonal changes and location. Because of the irregularities, sundial time is not very useful in modern times. Simply driving a short distance East or West would significantly change the time of day. For this reason, time of day is normally based upon a mean solar day.

Even though time of day is regular, the time of day is still related to a mean Sun and is therefore relative to an individual's position. For example, noon is when the Sun is "directly overhead." But when the Sun is "directly overhead" for someone in Boston, it is certainly not for someone in Los Angeles, nor for someone only a few hundred miles East or West of Boston. To circumvent problems relative to location, the world is divided up into 24 time zones. Each time zone is roughly 15° wide although allowances are made to accommodate country or other boundaries. The continental United States is divided into 4 time zones with very irregular boundaries as can be seen from a map which shows the time zones. Time within a zone is called Standard Time which is based on the local mean time (LMT) of that zone's standard meridian of longitude. (In keeping with Sidgwick's *Amateur Astronomer's Handbook*, Standard Time will not be differentiated from LMT.) All individuals within the same time zone agree upon the time even though the relative position of the mean Sun may differ for two different observers.

Time zones are measured relative to Greenwich, England. Geographical longitude is also measured from Greenwich, England. The local mean time for Greenwich is called the Greenwich Mean Time (GMT). Most astronomical calculations are based upon GMT which will be explained shortly. Beginning with Greenwich (0° longitude) being in the middle of the GMT time zone, the Earth is divided into the 24 time zones that are about 15° wide in longitude. Each time zone differs from the adjacent time zone by 1^h. Time zones West of Greenwich are earlier than GMT while time zones East of Greenwich are later than GMT. For instance, the Eastern Standard Time Zone of the United States is 5 time zones away from Greenwich. Hence, when an observer in the Eastern Standard Time thinks it is noon, an observer in Greenwich thinks it is 5:00 p.m.

The definition for GMT given here should not be confused with Greenwich Mean Astronomical Time (GMAT) which is reckoned from noon rather than midnight. Technically, GMT should be measured from the transit of a mean Sun through upper culmination and should thus be measured from noon instead of midnight. However, in accordance with common usage, GMT will be used synonymously with Universal Time (UT).

To review what has been said so far, the time on one's wristwatch is called his Standard Time which is based on the zone's local mean time (LMT). All observers within the same time zone agree upon time regardless of the position of the mean Sun. The world is divided up into time zones starting from Greenwich, England. The LMT for Greenwich, England is called the GMT and will be used in this book synonymously with Universal Time (UT). Within the continental United States, there are 4 time zones. These time zones are, from East to West: Eastern Standard Time Zone (EST), Central Standard Time Zone (CST), Mountain Standard Time Zone (MST), and Pacific Standard Time Zone (PST). Adjacent time zones are 1^h apart (15° longitude) with EST being 5 time zones away from Greenwich, England.

Some countries, or states, add or subtract an hour to the Standard Time depending upon the season. This is called daylight saving time (DST). Most areas of the United States have adopted DST and add an hour during the spring and subtract an hour during the fall. (This can be remembered by the phrase, *spring up, fall back*, which describes whether to set the clock ahead or back.) Remember to account for DST during time of day conversions.

Earlier it was stated that mean solar time is used to avoid the irregularities caused by the Earth's orbit. How-

Figure 4.3: *The Greenwich Observatory. Note that there is a ball resting on the left cupola of the main building. This ball could be seen by ships in the nearby harbor. The ball would be drawn up the pole and then lowered so that at precisely noon it touched the cupola. Ships then had a precise method to set their clocks which were needed for navigation.*

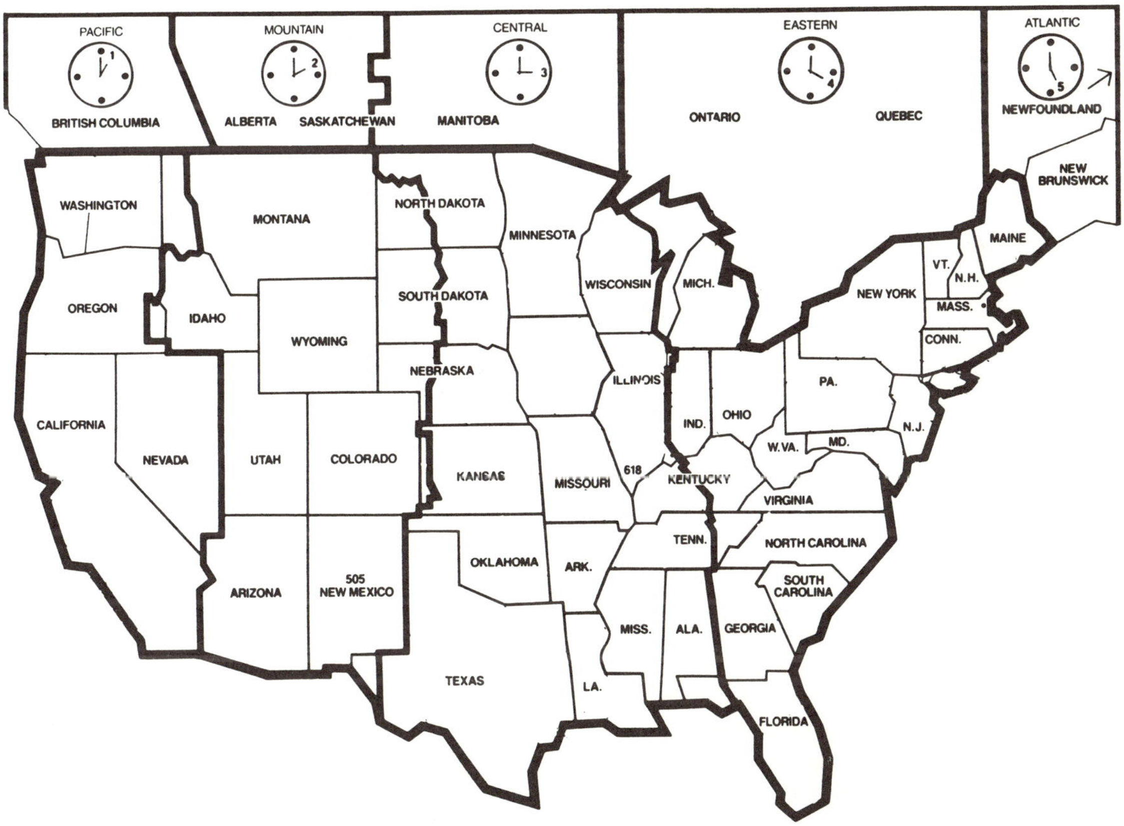

Figure 4.4: *This time zone map of the United States and Canada shows that political boundaries often influence how we keep time; otherwise the North–South dividing lines would be straight!*

ever, the mean solar time is still based upon the Earth's rotation about its own axis. This rotation is not uniform either so that the mean solar day is not regular in length. Astronomers have defined a system of time measurement called Ephemeris Time (ET) which accounts for this irregularity in the Earth's rotation. Ephemeris Time has been chosen to coincide with GMT as closely as possible. For most of the remainder of this book, it will be assumed that ET = GMT = UT. The error incurred can be measured for past dates, but only predicted for the future. However, the error in assuming that ET = GMT is unlikely to be more than a few minutes for this century. In January 1980 the error was slightly more than 50 seconds. For the purposes of this book, the accuracy is sufficient.

Just as the definition of a day can be tied to the stars (sidereal day), time can also be measured by the stars. Such time is called sidereal or "star" time. The sidereal time for two observers also differs depending upon their respective locations. Sidereal time at Greenwich is called Greenwich Sidereal Time (GST) while sidereal time for a specific observer is called Local Sidereal Time (LST). Two observers within the same time zone will not agree upon the LST unless they are in the same location. The reason for this is that a star is in a different place in the sky for observers at two different locations.

Because of wobbling in the Earth's rotation, sidereal time is not uniform just as mean solar time is not uniform. Astronomers have defined a system of measuring sidereal time, called mean sidereal time, to account for this wobbling in the Earth's axis. What has been described so far is called apparent sidereal time. The difference between apparent sidereal time and mean sidereal time is only a few seconds and will be ignored henceforth. Sidereal time will really refer to apparent sidereal time but will be assumed to equal mean sidereal time.

It is now appropriate to describe how the time conversions are done. For the moment concentrate more on *how* to do them rather than *why*. Generally speaking, most of the routines in later chapters will ask for the LMT for a given location. Once the LMT is known, it is usually desirable to convert it to the LST. The process involved is to convert LMT to GMT, GMT to GST, and finally convert GST to LST. Converting LST to LMT works by just reversing this process. The conversions below will assume that the time is entered in HMS format (24 hour clock).

In the time conversion procedures below, it is often necessary to compute a time zone adjustment. The zone adjustment depends upon which time zone that the observer is located in. Table 4.1 gives the time zone adjustments in hours for each of the time zones in the continental United States. If the time zone is unknown,

Time Zone	Adjustment
EST	5 hours
CST	6 hours
MST	7 hours
PST	8 hours

Table 4.1 Time Zone Adjustments. *The continental United States is divided into four time zones each 1 hour apart.*

a time zone adjustment can be calculated as

$$Adjustment = \texttt{FIX}\left(\frac{\psi}{15}\right) \tag{4.5.1}$$

where ψ represents the observer's geographic longitude. Longitude, ψ, is measured in decimal degrees and is positive for locations West of Greenwich and negative for longitudes East of Greenwich. For example, the zone adjustment for 30°W is 2^h while the zone adjustment for 45°E is -3^h.

4.6 LMT to GMT (UT)

Convert $18^h0^m0^s$ LMT to GMT for an observer in the Eastern Standard Time Zone. Assume that this is not daylight saving time.

1. Convert *LMT* to decimal format.
   (Ans: $LMT = 18.0^h$)
2. If necessary, adjust for daylight saving time.
   If the *LMT* given is on *DST*, subtract 1^h.
   Otherwise do nothing in this step.
   (Ans: no adjustment needed)
3. Using equation 4.5.1, calculate a time zone adjustment.
   (Ans: $Adjustment = 5^h$)
4. Add the time zone adjustment to the result of step 2.
   (Ans: 23^h)
5. If (Result of step 4) < 0, add 24^h.
   If (Result of step 4) > 24, subtract 24^h.
   (Note that if 24^h must be added to step 4, the resulting time is on the previous date while if 24^h must be subtracted, the resulting time is for the next day.)
   (Ans: no adjustment)
6. Convert the result of step 5 to HMS format.
   (Ans: $GMT = 23^h0^m0^s$)

4.7 GMT (UT) to LMT

Convert $23^h30^m0^s$ GMT to LMT for an individual within the Eastern Standard Time zone and assume daylight saving time.

1. Convert the GMT time to decimal format.
(Ans: $GMT = 23.5^h$)
2. Using equation 4.5.1, calculate a time zone adjustment.
(Ans: $Adjustment = 5^h$)
3. Subtract the time zone adjustment from the result of step 1.
(Ans: 18.5^h)
4. If (Result of step 3) > 24, subtract 24^h.
If (Result of step 3) < 0, add 24^h.
(Note that adding 24^h means that the LMT is for the next day while subtracting 24^h means that the LMT is for the previous day.)
(Ans: no adjustment)
5. If the observer's location is on daylight saving time, add 1^h to step 4.
(Ans: 19.5^h)
6. Convert the LMT into HMS format.
(Ans: $LMT = 19^h30^m0^s$)

4.8 GMT (UT) to GST

To convert GMT to GST, the date must be known. Convert $23^h30^m0^s$ GMT to GST for February 7, 1985.

1. Convert the given date to its Julian day number (do not express the time of day as a fractional part of the day).
(Ans: $JD = 2,446,103.5$)
2. Calculate the Julian day number for January 0.0 of the given year. Let this Julian date be JD_0.
(Ans: $JD_0 = 2,446,065.5$)
3. Subtract step 2 from step 1 to get the number of elapsed days into the year. Let $Days$ be this number.
(Ans: $Days = 38$)
4. Compute a temporary value
$$T = \frac{JD_0 - 2,415,020.0}{36,525}.$$
(Ans: $T = 0.849979$).
5. Compute
$$R = 6.6460656 + 2,400.051262 * T + 0.00002581 * T^2.$$
(Ans: $R = 2,046.640375$)
6. $B = 24 - R + 24 * (Year - 1900)$.
(Ans: $B = 17.359625$)
7. $T_0 = 0.0657098 * Days - B$.
(Ans: $T_0 = -14.862653$)
8. Convert the GMT given into decimal format.
(Ans: $GMT = 23.5^h$)
9. $GST = T_0 + 1.002738 * GMT$.
(Ans: $GST = 8.701690^h$)
10. If (GST from previous step) < 0, add 24^h.
If (GST from previous step > 24, subtract 24^h.
(Ans: no adjustment)
11. Convert the GST from the previous step into the HMS format.
(Ans: $GST = 8^h42^m6^s$)

4.9 GST to GMT (UT)

Converting GST to GMT requires that the date be known. Calculate the GMT for $2^h3^m41^s$ GST on February 7, 1985.

1. Convert the given date (at 0^h) to its Julian day number.
(Ans: $JD = 2,446,103.5$)
2. Calculate the Julian date for January 0.0 of the given year. Call this Julian day number JD_0.
(Ans: $JD_0 = 2,446,065.5$)
3. Subtract step 1 from step 2 to get the number of days into the year. Call this number $Days$.
(Ans: $Days = 38$)
4. Calculate a temporary value
$$T = \frac{JD_0 - 2,415,020.0}{36,525}.$$
(Ans: $T = 0.849979$)
5. Calculate
$$R = 6.6460656 + 2,400.051262 * T + 0.00002581 * T^2.$$
(Ans: $R = 2,046.640375$)
6. $B = 24 - R + 24 * (Year - 1900)$.
(Ans: $B = 17.359625$)
7. $T_0 = 0.0657098 * Days - B$.
(Ans: $T_0 = -14.862653$)
8. If (Result of step 7) < 0, add 24^h.
If (Result of step 7) > 24, subtract 24^h.
(Ans: $T_0 = 9.137347$)
9. Convert the GST given to decimal format.
(Ans: $GST = 2.061389^h$)

10. $A = GST - T_0$.
    (Ans: $A = -7.075958$)
11. If $A < 0$, add 24^h.
    (Ans: $A = 16.924042$)
12. $GMT = 0.997270 * A$.
    (Ans: $GMT = 16.877839^h$)
13. Convert GMT to HMS format.
    (Ans: $GMT = 16^h52^m40^s$)

Note that steps 1–7 are identical to converting GMT to GST.

4.10 GST to LST

Converting GST to LST requires knowing the observer's longitude. Assume that GST is $2^h3^m41^s$ for an observer at longitude 40°W. Calculate the corresponding LST.

1. Convert the GST to decimal format.
   (Ans: $GST = 2.061389^h$)
2. Calculate a zone adjustment.

$$Adjust = \frac{\psi}{15}.$$

   Note that this adjustment is almost the same as the time zone adjustment in equation 4.5.1.
   (Ans: $Adjust = 2.666667$)
3. $LST = GST - Adjust$.
   (Ans: $LST = -0.605278$)
4. If $LST > 24$, subtract 24^h.
   If $LST < 0$, add 24^h.
   (Ans: $LST = 23.394722$)
5. Convert LST to the HMS format.
   (Ans: $LST = 23^h23^m41^s$)

Notice that the local sidereal time is simply the Greenwich sidereal time with an adjustment that takes into account the observer's longitude. This adjustment is the difference, expressed in hours, in the observer's longitude and Greenwich, England.

4.11 LST to GST

Converting LST to GST is very similar to converting GST to LST. It requires knowing the observer's location. Assume that an observer at longitude 50°E calculates LST as $23^h23^m41^s$. Convert LST to GST.

1. Convert LST to decimal format.
   (Ans: $LST = 23.394722^h$)
2. Calculate a time zone adjustment.

$$Adjust = \frac{\psi}{15}.$$

   (Ans: $Adjust = -3.333333$)
3. $GST = LST + Adjust$.
   (Ans: $GST = 20.061389$)
4. If $GST < 0$, add 24^h.
   If $GST > 24$, subtract 24^h.
   (Ans: no adjustment)
5. Convert GST to HMS format.
   (Ans: $GST = 20^h3^m41^s$)

4.12 About the Program

The program TIME contains the routines to do all of the time and date conversions described in this chapter. Since most versions of `BASIC` do not include a suitable `FRAC` function, a `BASIC` function is defined to make up for this deficiency. The function, `FNFRAC`, is defined as

$$\texttt{FNFRAC}(x) = \texttt{ABS}(x - \texttt{FIX}(x)).$$

The program TIME performs the appropriate conversions required for time zones within the continental United States. For a location outside this area, type a `RETURN` to use the longitude adjustment given in equation 4.5.1.

Several of the subroutines presented in this chapter will be used in later chapters. As a general practice, all of the programs have `REM` blocks at the beginning of the subroutine to indicate what variables are assumed to be set before the subroutine is invoked, what variables are used inside the subroutine, what other subroutines are called, and what the result variables are. Should you decide to modify any of the subroutines, it is important to make sure that the proper variables are set up before making the subroutine calls else the results returned will be invalid.

4.13 Sample Problems

Following are several exercises for this chapter.

1. Is 1984 a leap year?
   (Ans: yes)
2. Is 1954 a leap year?
   (Ans: no)
3. Is 2000 a leap year?
   (Ans: yes)
4. Is 1900 a leap year?
   (Ans: no)
5. Convert midnight November 19, 1954 to its Julian day number.
   (Ans: 2,435,065.5)
6. Convert midnight May 10, 2 B.C. to its Julian date.
   (Ans: 1,720,456.5)

7. Convert 2,369,915.5 to its corresponding calendar date.
   (Ans: 7/4/1776)
8. Convert 1,721,781.5 to its corresponding calendar date.
   (Ans: 12/25/1)
9. What day of the week does 7/4/1776 fall on?
   (Ans: Thursday)
10. What day of the week does 1/20/1985 fall on?
    (Ans: Sunday)
11. What is the day number for 10/31/1984?
    (Ans: 305)
12. If the date is 250 days into 1900, what is the date?
    (Ans: 9/7/1900)
13. Assume that the date is 12/12/1984 and that an observer is in the Eastern Standard Time Zone at 45°W longitude. Assume that it is not daylight saving time. If LMT is $20^h0^m0^s$, what are the corresponding GMT, GST, and LST times?
    (Ans: $GMT = 1^h0^m0^s$ (next day!),
    $GST = 6^h27^m37^s$ (12/13/1984), and
    $LST = 3^h27^m37^s$ (12/13/1984))
14. Assume that the date is 7/5/1900 for an observer in the Mountain Standard Time Zone at 60°W longitude. Assume that it is daylight saving time. If LST is $8^h53^m4^s$, what are the corresponding GST, GMT, and LMT times?
    (Ans: $GST = 12^h53^m4^s$, $GMT = 18^h0^m0^s$,
    and $LMT = 12^h0^m0^s$)

Chapter 5

Coordinate System Conversions

Only one major piece of information is missing before being able to calculate the location of a star or planet. That missing piece is the ability to describe the location of an object on a sphere. In Chapter 4, the time element of positional astronomy was discussed. The tools presented in that chapter account for differences in time measurements and differences in location on the Earth. The present chapter will discuss techniques for locating an object on a sphere using some coordinate system and techniques for converting between different coordinate systems. These techniques will be applied over and over in the succeeding chapters which actually compute the location of a celestial object.

This chapter will describe five different coordinate systems that are used in astronomy to describe the location of an object. The first system to be described is the latitude-longitude system which is used to locate an object on the surface of the Earth. This system has already been used in Chapter 4 but will be treated more thoroughly here. A second system, the horizon coordinate system, is an easy method for determining the location of an object in the sky. However, the coordinates of an object expressed in the horizon coordinate system are constantly changing as the Earth rotates and are different for observers at different locations. To avoid such problems, a third coordinate system, the equatorial coordinate system, is defined so that the position of an object is the same regardless of the observer's location on the Earth. Because the stars are so far away, the equatorial coordinates so defined change very slowly with time. For most practical purposes, the stars are considered stationary and slight adjustments in the equatorial coordinates to allow for motion of the stars are made only for extreme accuracy. Planets, on the other hand, move much more rapidly in their orbits so that their corresponding equatorial coordinates change daily. Calculations for planetary positions are generally done in the ecliptic coordinate system for convenience and then converted to equatorial coordinates. Finally, calculations involving objects within the Milky Way Galaxy are done within the galactic coordinate system and then converted to the equatorial coordinate system. For completeness, the galactic coordinate system will be described in this chapter but will not be used further.

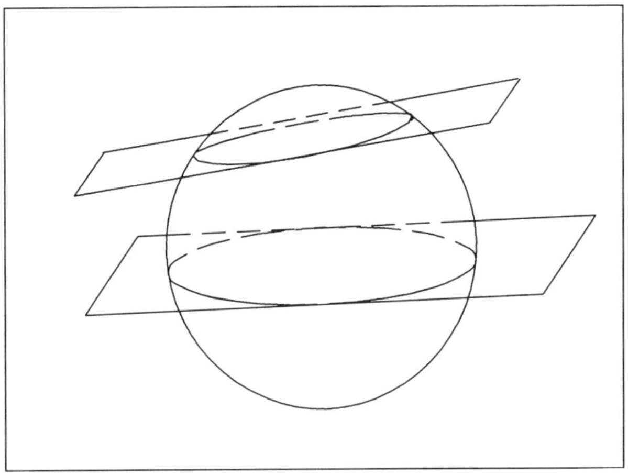

Figure 5.1: *Cutting a sphere as shown here can result in different sized circles. When a circle is formed that has the same diameter as the sphere itself it is called a great circle.*

5.1 Locating Objects on a Sphere

The latitude-longitude system is the most familiar technique for locating an object on a sphere. Several definitions are needed to properly describe this system. Visualize a sphere for a moment. A line which passes through the center of the sphere is equal to the diameter of the sphere. In the case of the Earth, a line passing from the North Pole to the South Pole passes through the center of the Earth and is the same length as the Earth's diameter. Besides being a diameter, this particular line is also the Earth's axis of rotation. (Strictly speaking, the Earth is not a sphere. However, for our purposes it will be considered as being perfectly spherical in shape.)

Returning to our imaginary sphere, visualize passing a sheet of paper through the sphere. Figure 5.1 illustrates an example. No matter how the sheet is passed through the sphere, a circle is formed by the intersection

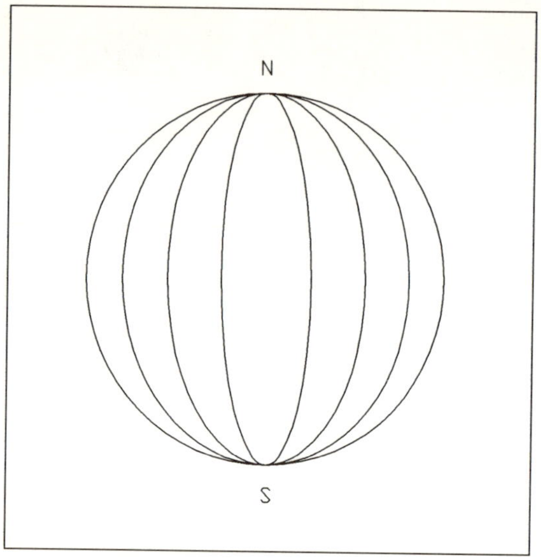

Figure 5.2: *Great circles which pass through both the North and South Pole are called meridians.*

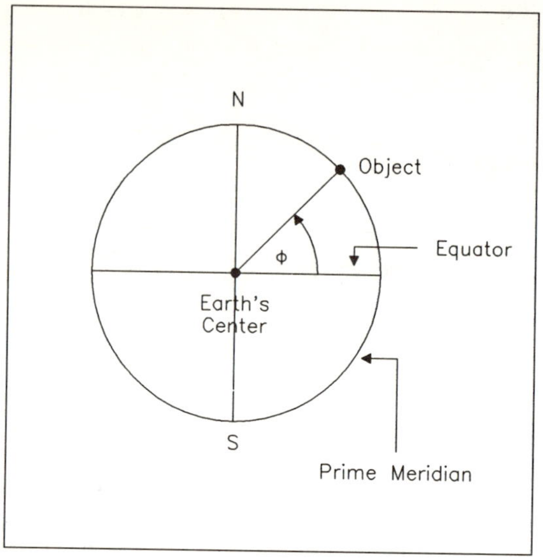

Figure 5.3: Measuring Latitude. *Latitude is a measure of an object's distance from the equator.*

of the paper with the sphere. But some of the circles formed are larger than others. When the circle formed has the same diameter as the sphere itself, the circle is called a great circle. In the case of the Earth, a circle which goes through both the North and South Poles is a great circle. The equator is another example of a great circle. There are an infinite number of great circles that can be drawn on a sphere.

Great circles which pass through both the North and South Pole of the Earth are called meridians. Figure 5.2 shows meridians drawn on the surface of the Earth. The meridian which passes through Greenwich, England is called the prime meridian. Actually, the prime meridian is only half of the great circle indicated. The prime meridian is the half circle which starts at the North Pole, goes through Greenwich, England, and terminates at the South Pole.

Location on the prime meridian can be easily described by specifying, in degrees, how far an object is above or below the equator. The prime meridian forms a semi-circle whose center is the center of the Earth. Using the equator as the starting point, the location of an object on the prime meridian can be described by measuring the angle that is formed by a line from the object to the center and a line from the center to the equator. Figure 5.3 illustrates this angle. This measurement is called the object's latitude and is represented by the symbol ϕ. Objects on the equator are at 0° latitude while objects above the equator are in the range of 0° to 90°N. Objects below the equator are in the range of 0° to 90°S. The North Pole is 90°N latitude while the South Pole is 90°S latitude. An object that is exactly halfway between the North Pole and the equator is at 45°N latitude.

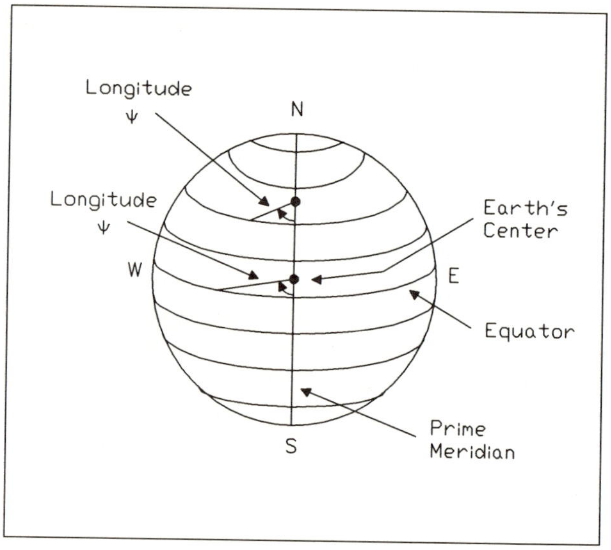

Figure 5.4: Measuring Longitude. *Longitude is a measure of an object's distance from a prime meridian.*

Obviously, not all objects of interest happen to fall on the prime meridian. Starting with the equator, a series of circles are drawn around the Earth such that every point on the circle has the same latitude. Figure 5.4 illustrates these circles. Notice that only the equator is a great circle. An object can thus be located by indicating how far, in degrees, it is away from the prime meridian. This measurement is called the longitude, designated by the symbol ψ, with the prime meridian being 0° longitude. From the prime meridian, longitudes are in the range 0° to 180°W or 0° to 180°E. Thus, latitude and longitude precisely describe any location on the Earth. Figure 5.5 shows the Earth with latitude and longitude circles.

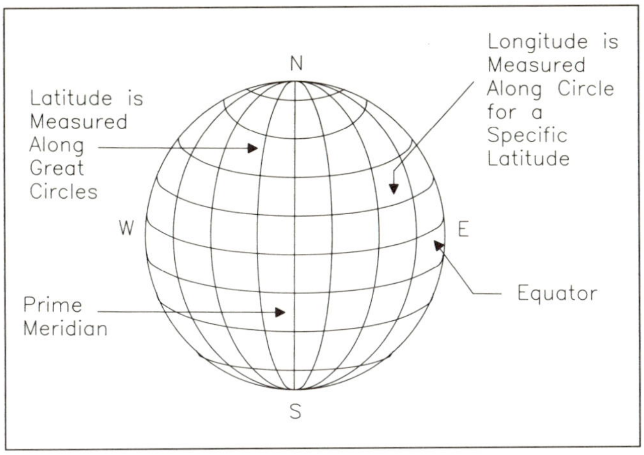

Figure 5.5: Longitude and Latitude Circles. *Latitude and longitude precisely specify an objects location.*

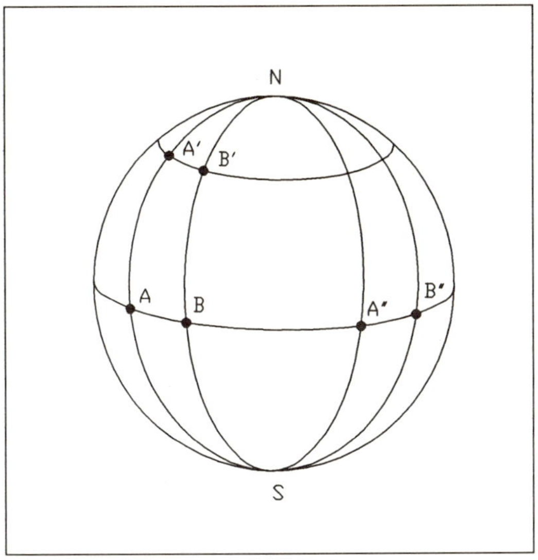

Figure 5.6: Locations with the the same latitude. *Changing latitude does not preserve distance between points.*

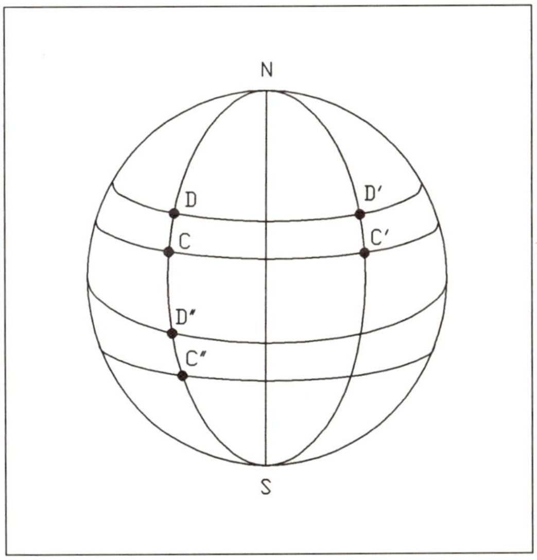

Figure 5.7: Locations with the same longitude. *Distance along a great circle is independent of the great circle's location.*

Longitude is sometimes measured in hours, minutes, and seconds. Since there are 360° in a circle, one hour corresponds to 15°. This book will assume that longitude (on the Earth) is measured in degrees.

Consider measuring the distance between two different locations. Assume that location A and B are at the same latitude but have different longitudes. Now keeping the difference in longitudes the same, visualize moving A and B to a different latitude. This situation is marked as A' and B' in Figure 5.6. The distance, in miles, between A' and B' is not the same distance as between A and B. Instead of changing latitudes, suppose that A and B are again moved such that the difference in their longitudes is the same as before but this time their latitudes are not changed. These new points are A'' and B'' in Figure 5.6. This time, the distance between A'' and B'' is exactly the same as between A and B. The reason for this is that distance measured along an arbitrary circle depends very much upon its latitude. Since A' and B' are at a different latitude than A and B, their distance is different. But A'' and B'' are at the same latitude as A and B and are thus being measured along a circle which is at the same latitude for both pairs of locations.

An analogous situation is pictured in Figure 5.7. Locations C and D are at the same longitude but have different latitudes. Keeping the difference in latitude the same but changing the longitude produces points C' and D'. The distance from C to D is the same as the distance from C' to D'. This is because distance is being measured along a great circle. Distance along a great circle is independent of where the great circle is located. C'' and D'' in Figure 5.7 are produced by keeping the longitude the same but moving to a new latitude such that the difference in latitude between C'' and D'' is the same as the difference in latitude between C and D. The distance between C'' and D'' is again the same as between C and D because distance is being measured along a great circle.

The importance of this discussion is that measuring distance between two points on a sphere must be done along a great circle which passes through both points. That is why routes flown by airplanes do not appear on a map as straight lines. Distance measured along a great circle is usually shorter than what we perceive on most maps as a straight line.

5.2 The Celestial Sphere

A convenient way to describe the location of stars is to assume that they are embedded on a large sphere with the Earth being at the center of that sphere. This sphere is called the celestial sphere. Recall that this idea can be traced back to at least Aristotle as explained in Chapter 1. By extending the plane of the Earth's equator until it intersects the celestial sphere, the celestial equator is formed. Extending the Earth's North Pole until it intersects the celestial sphere defines the location of the North Celestial Pole. The star Polaris, often called the North Star or Pole Star, is very close to the North Celestial Pole.

At this point, it is probably clear that the celestial sphere can also be divided into meridians in the same manner as the Earth. But how is a celestial prime meridian to be determined? At first glance, it might appear that the Earth's prime meridian could be extended outwards until it intersects the celestial sphere. But since the Earth rotates on its axis, this would have the undesirable effect of having a celestial prime meridian which changes with the time of day. Instead, astronomers have chosen a fixed point in the sky called the First Point of Aries, denoted by the symbol ♈, to play a role analogous to Greenwich, England. Using this point, a celestial prime meridian can be defined as a great circle (actually, only half a circle) which begins at the North Celestial Pole, passes through the First Point of Aries, and terminates at the South Celestial Pole. The First Point of Aries is also called the vernal equinox.

Meridians on the celestial sphere can be specified in a couple of ways. Celestial meridians can be defined as great circles which pass through the North and South Celestial Poles. Defined this way, celestial meridians do not vary with the time of day. Another way to define meridians is to define them relative to an observer on the Earth. Consider Figure 5.8. An observer is located at position O and his horizon is the circle $NESW$ where these letters represent compass directions. Notice that if the line NS is extended to intersect the celestial sphere, it will not define the North Celestial Pole. If the observer is at the Earth's equator, the line formed will be parallel to the line connecting the North and South Celestial Poles. Suppose P in Figure 5.8 is a star on the celestial sphere. A celestial meridian is then defined by the great circle NPS. Of course, relative to the stars this meridian will change with the time of day. But relative to the observer this meridian stays fixed and it is the stars which move. As a star appears to move and crosses a meridian, the star is said to transit or culminate.

From Figure 5.8, point Z is the point that is directly overhead the observer located at position O. This point is called the observer's zenith. The point on the celestial sphere which is directly beneath the observer is called the nadir. The great circle NZS in Figure 5.8 is called the observer's meridian. Recall from Chapter 4 that a day is defined as the interval between two successive transits of a mean Sun (for mean solar day) or a fixed star (sidereal day) across an observer's meridian. When no specific meridian is mentioned, transit (or culminate) is taken to mean with respect to the observer's meridian.

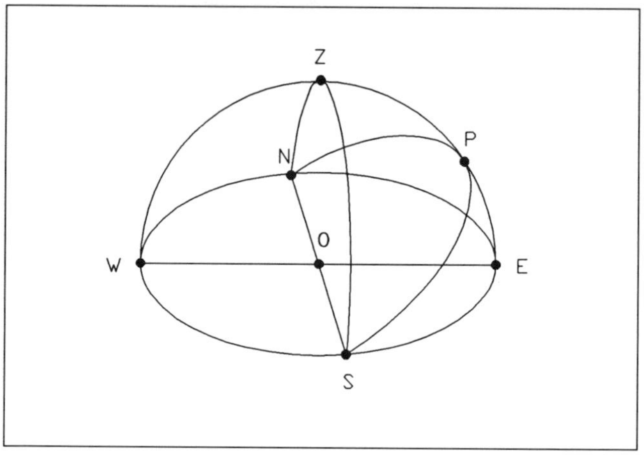

Figure 5.8: An observer's horizon. *Celestial meridians defined with respect to an observer vary with the time of day.*

Meridians defined relative to an Earthbound observer will appear to move as the Earth rotates when viewed from a location outside the earth. Furthermore, no two Earthbound observers can have the same observer's meridian unless they are at the same location.

5.3 Orbital Elements

Determining the position of a planet, the Sun, or the Moon is not quite as simple as determining the location of a star. For one thing, the stars can be considered as fixed unless extreme accuracy is required. On the other hand, objects within the solar system move in orbits that are easily perceivable from the Earth. Actually calculating the position of an object within the solar system will not be covered until Chapters 7, 8, and 9, but at this point it is appropriate to discuss orbital elements which must be considered to locate an object within the solar system.

Kepler's laws explain that orbits are elliptical in shape. It is possible for an orbit to be a perfect circle (mathematically a circle is a special case of an ellipse), but none of the objects within the solar system have a circular orbit. It is often convenient, however, to

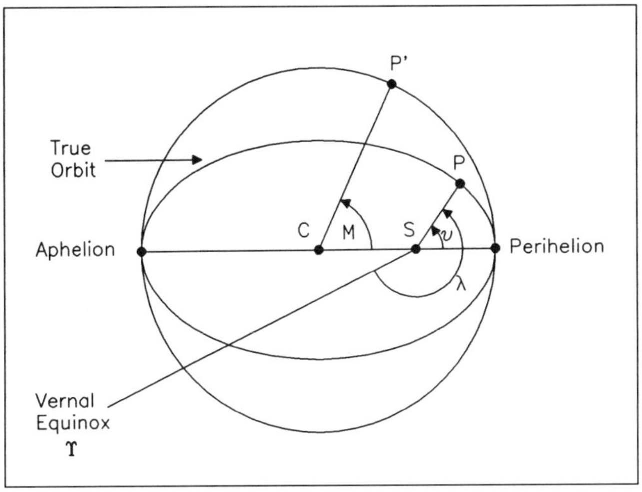

Figure 5.9: Elliptical orbit. *A true elliptical orbit can be superimposed on a sphere.*

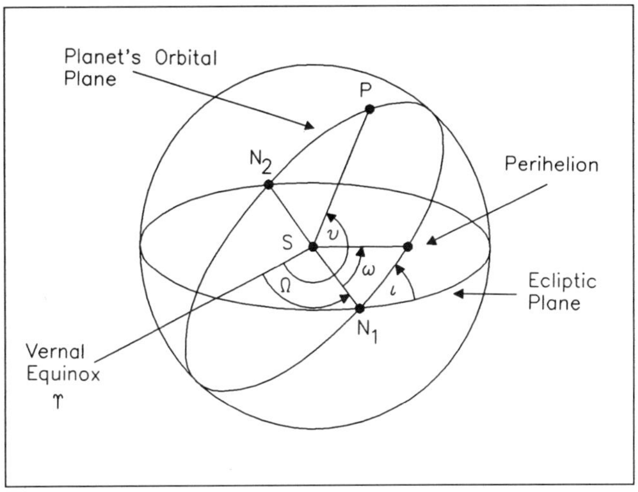

Figure 5.10: Orbital plane. *Calculating the location of a planet is complicated by the fact that the planet's orbit does not lie in the Earth-Sun plane.*

assume that orbits are perfect circles and hence move in great circles on some sphere. Figure 5.9 shows how the true elliptical orbit can be superimposed on a sphere. According to Kepler's laws, the Sun, S, is at one focus of the ellipse for planet P's orbit. When the planet is closest to the Sun, it is said to be at perihelion. When the planet is farthest away from the Sun, it is said to be at aphelion. Both of these locations are indicated in Figure 5.9. A circle whose diameter is the aphelion-perihelion distance is also shown with it's center marked by C. This circular orbit is called the planet's mean orbit. The true position of the planet is at P while it's mean position is at P'. That is, P' indicates where the planet would be if it did indeed follow a circular orbit.

There are three angles indicated in Figure 5.9. The angle labeled M is measured from a line through C and perihelion to a line extending from C to P', the planet's mean position. This angle is called the mean anomaly. The angle labeled υ is called the true anomaly and indicates the true position of the planet. Angle λ is measured from the vernal equinox to the planet's true position and is called the ecliptic longitude of the planet. This will be explained more fully below when the ecliptic coordinate system is defined.

The plane containing the Earth and Sun is called the ecliptic plane. Earth's orbit lies entirely within the ecliptic plane. However, planetary orbits do not lie within the ecliptic plane. Figure 5.10 illustrates a sphere containing the ecliptic plane as its "equator." The orbit for a planet is also shown and labeled as the planet's orbital plane. The Sun, S, is located in the center of the sphere with the point of perihelion, location of planet P, vernal equinox ♈ (First Point of Aries), and the true anomaly υ also labeled. When the planet changes from being below the ecliptic plane to being above the ecliptic plane, it has passed through the ascending node N_1. N_2 is the descending node and is the point at which the planet goes below the ecliptic plane. Angle Ω in Figure 5.10 is measured from the vernal equinox to the ascending node and is known as the longitude of the ascending node. Angle ω , measured from the ascending node to the point of perihelion, is called the argument of perihelion. Angle ι is the angle of inclination and measures the inclination between the ecliptic plane and the planet's orbital plane.

The angles shown in Figures 5.9 and 5.10 can be used to completely describe the orbital characteristics of any object with an elliptical orbit. The chapters which lie ahead will return to these angles to describe the orbits of the Sun, Moon, and planets and will use them to calculate positions for these objects for any given date, time, and position on the Earth.

5.4 Coordinate System Conversions

Before looking at how the coordinate system conversions are done, it is necessary to digress for a moment and consider some properties of the trigonometric functions. In many of the coordinate system conversions, inverse trigonometric functions are required to find the desired angle. Trigonometric functions are defined over the entire range of real numbers and, excepting the tangent and cotangent functions, always produce as a result a real number between -1 and $+1$ inclusive. This leads to

y	x	Adjustment
+	+	0
+	−	180
−	+	360
−	−	180

Table 5.1 Angle adjustment for Arctangent. *Computing the arctangent requires an adjustment to place the answer in the proper quadrant.*

an ambiguity when inverse trigonometric functions are involved. For example, the tangent of 45° is 1 which is the same value as for the tangent of 225°. What then is the inverse tangent of 1 to be, 45° or 225°? To determine the correct answer, inverse trigonometric functions will have to be considered on a case by case basis in the context of the problem being solved.

By convention, the inverse cosine function ($\cos^{-1}$) returns angles between 0° and 180° while the inverse sine ($\sin^{-1}$) and inverse tangent ($\tan^{-1}$) return angles between −90° and +90°. The inverse sine and inverse cosine functions will rarely cause a problem in the conversions below, but the inverse tangent function does pose an added complexity. When the inverse tangent function is required and the resulting angle is supposed to be in the range of 0° to 360°, the argument to the inverse tangent function will be given in the form y/x. (A function's argument is the value to which the function is to be applied. Thus, in the expression $\sin 45°$, 45° is the argument for the sin function.) The correct angle will depend upon the numeric sign of y and x. There are four possible cases to consider as summarized in Table 5.1. Remember, however, that the adjustment is applied only if the resulting angle is to be in the range [0°, 360°] instead of [−90°, +90°].

As an example, suppose that $y = 5$, $x = -2$, and the desired angle θ is to be in the range 0° to 360°. Now

$$\theta = \tan^{-1}\left(\frac{5}{-2}\right) = -68.1986°.$$

Table 5.1 indicates that 180° must be added to θ to give an angle in the correct quadrant. Therefore, the correct angle is

$$\theta = -68.1986° + 180° = 111.8014°.$$

It is important to remember that when an inverse trigonometric function is needed, an angle adjustment may or may not be needed. An angle adjustment will be needed for the inverse sine and inverse tangent functions if the desired angle could fall outside the range of −90° to +90°. An adjustment will be needed for the inverse cosine function if the desired angle could fall outside the range of 0° to 180°. Failure to consider the inverse trigonometric functions on a case by case basis in the context of the problem being solved can produce an incorrect result.

It is now time to discuss the coordinate systems used to specify the location of an object. Four such systems will be described: horizon, equatorial, ecliptic, and galactic. The horizon and equatorial coordinate systems have already been mentioned in Chapter 3 when telescope mountings were discussed. The altazimuth mounting allows a telescope to be positioned by specifying a location in the horizon coordinate system. Likewise, the equatorial mounting allows a telescope to be pointed by specifying a location in the equatorial coordinate system.

5.5 Equatorial Coordinate System

The equatorial coordinate system may be the most widely used astronomical coordinate system. Star atlases and star catalogues generally give the location of stars, nebulae, quasars, etc. in equatorial coordinates. Monthly astronomy journals, such as *Sky & Telescope*, frequently list the location of solar system objects by equatorial coordinates. The reason is of course that equatorial coordinates describe a location that is independent of time or an observer's location. Moreover, the procedures below will describe how to convert any other coordinate system to the equatorial system. Figure 5.11 shows the process. As an example, to convert galactic coordinates to horizon coordinates, convert the galactic coordinates to equatorial and then convert the equatorial coordinates to horizon coordinates.

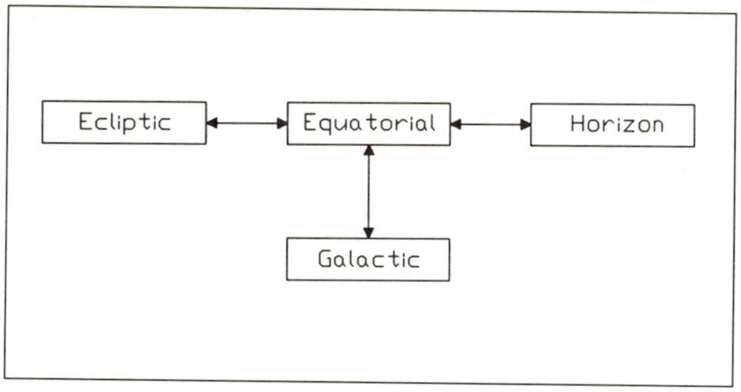

Figure 5.11: *Coordinate system conversions are simplified by using the equatorial coordinate system as an intermediate step.*

The equatorial coordinate system is similar to the latitude-longitude method for locating a position on the Earth. The celestial equator plays the same role as the Earth's equator while the First Point of Aries plays the role of Greenwich, England. Instead of using longitude and latitude, the equatorial coordinate system uses measurements called declination (δ) and right ascension (α). Declination is similar to latitude in that it indicates how far an object is from the equator. Declination is measured in degrees and falls within the range $-90°$ to $+90°$ with positive angles indicating locations North of the celestial equator and negative angles indicating locations South of the celestial equator. Right ascension and longitude are similar, but right ascension is normally expressed in HMS format and is in the range of $[0^h, 24^h]$ while longitude is normally expressed as an angle between $0°$ and $180°$ with a designation of East or West. When expressed in decimal format, right ascension can be converted to degrees by multiplying by 15 and thus make the right ascension measure even more similar to longitude.

Both the declination and right ascension for a given star are fixed and do not vary with the time of day or location. But instead of using right ascension to play the role of "celestial longitude," another measurement called the hour angle (H) is frequently used. While the right ascension is really an angle measurement, a star's hour angle is a time measurement. The hour angle for a star is a measure of how long it has been since the star crossed the observer's meridian. Note that because of the way that the hour angle is defined, it varies with time of day and the observer's location. Notice also that the hour angle is time relative to a sidereal day, not a mean solar day. A star will transit the observer's meridian again after 24 sidereal hours have elapsed at which time the star's hour angle is 0^h. Moreover, a star will transit when the local sidereal time (LST) is equal to the right ascension.

Converting between hour angle and right ascension is straightforward since the hour angle is defined to be the difference in the LST and right ascension. That is,

$$H = LST - \alpha. \tag{5.5.1}$$

This equation again points out that a star will transit when the LST is equal to the right ascension. Recall from Chapter 4 that calculating the LST involves the observer's location. Thus, as would be expected, the hour angle for a star will indeed vary with the observer's location.

As an example, consider a star whose right ascension is $3^h24^m6^s$. Suppose the LST is 18^h. Calculate the corresponding hour angle. The required steps are:

1. Convert LST to decimal format.
   (Ans: $LST = 18^h$)
2. Convert the right ascension to decimal format.
   (Ans: $\alpha = 3.401667^h$)
3. $H = LST - \alpha$.
   (Ans: $H = 14.598333^h$)
4. If $H < 0^h$, add 24^h.
   (Ans: no adjustment)
5. Convert H to HMS format.
   (Ans: $H = 14^h35^m54^s$)

This star will culminate when the hour angle is $3^h24^m6^s$.

The reverse process is no more difficult. Equation 5.5.1 can be rewritten as

$$\alpha = LST - H \tag{5.5.2}$$

to compute α given LST and H. Assume a planet's hour angle is 1^h15^m and the LST is 21^h. Find the corresponding right ascension.

1. Convert the LST to decimal format.
   (Ans: $LST = 21.0^h$)
2. Convert H to decimal format.
   (Ans: $H = 1.25^h$)
3. $\alpha = LST - H$.
   (Ans: $\alpha = 19.75^h$)
4. If $\alpha < 0^h$, add 24^h.
   (Ans: no adjustment)
5. Convert α to HMS format.
   (Ans: $\alpha = 19^h45^m0^s$)

The subroutine for doing these two conversions in the program COORD will start with the GMT rather than the LST as done here.

Telescopes with equatorial mounts contain two graduated circles, called setting circles, to assist in positioning the telescope. One setting circle measures declination but the other setting circle is usually for measuring the hour angle rather than the right ascension, although some telescopes have direct reading right ascension circles. This is especially convenient for mountings with a clock drive (ticking off sidereal time, not mean solar time). Thus, it is important to be able to calculate the hour angle.

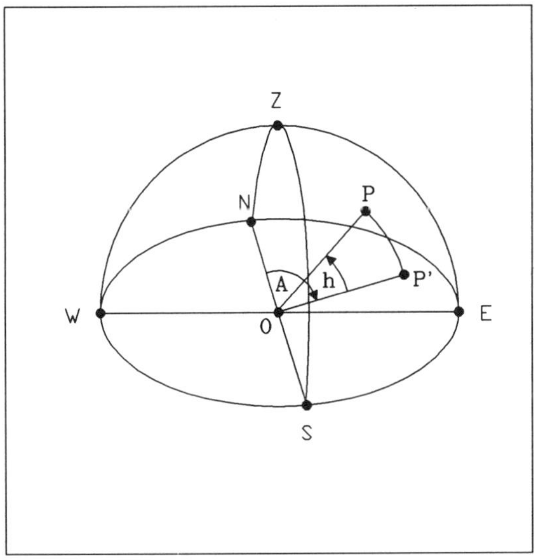

Figure 5.12: *The horizon coordinate system is easy to use, but it varies with location and time of day.*

5.6 Horizon Coordinate System

Horizon coordinates are based upon two angles which tell how far around an object is along the observer's horizon and how far up from the ground it is. Figure 5.12 shows an observer and his horizon. P is an object in the sky whose coordinates are desired. P' is the projection of P onto the plane of the observer. Angle NOP' is called the azimuth and measures how far object P is along the observer's horizon. Azimuth is measured from the North and is the range of 0° to 360°. The remaining angle, $\angle POP'$, is called the altitude and measures how high up the object is. Altitude ranges from −90° to +90°. Positive angles indicate objects above the horizon while negative angles indicate objects below the horizon. The symbol A represents azimuth and the symbol h represent altitude.

There are four equations which relate horizon and equatorial coordinates. The first two express equatorial coordinates in terms of horizon coordinates while the second two express horizon coordinates in terms of equatorial coordinates.

$$\sin\delta = \sin h * \sin\phi + \cos h * \cos\phi * \cos A \quad (5.6.1)$$

$$\cos H = \frac{\sin h - \sin\phi * \sin\delta}{\cos\phi * \cos\delta} \quad (5.6.2)$$

$$\sin h = \sin\delta * \sin\phi + \cos\delta * \cos\phi * \cos H \quad (5.6.3)$$

$$\cos A = \frac{\sin\delta - \sin\phi * \sin h}{\cos\phi * \cos h}. \quad (5.6.4)$$

All measurements in these equations are angles. (That is, right ascension α is converted to an angle before being used in an equation.)

Consider converting a horizon location to the equatorial coordinate system. Let us use altitude 40°, azimuth 115° as an example. Assume the observer is at 38°N latitude.

1. Convert h to decimal format.
   (Ans: $h = 40.0°$)
2. Convert A to decimal format.
   (Ans: $A = 115.0°$)
3. Compute
   $$\sin\delta = \sin h * \sin\phi + \cos h * \cos\phi * \cos A.$$
   where ϕ is the observer's latitude.
   (Ans: $\sin\delta = 0.140625$)
4. Take the inverse sine of step 3 to get δ.
   (Ans: $\delta = 8.084040°$)
5. $T = \sin h - \sin\phi * \sin\delta$.
   (Ans: $T = 0.556210$)
6. Compute
   $$\cos H = \frac{T}{\cos\phi * \cos\delta}.$$
   (Ans: $\cos H = 0.712925$)
7. Take the inverse cosine to get H.
   (Ans: $H = 44.526593°$)
8. Compute $\sin A$. If $\sin A > 0$, then $H = 360° - H$.
   (Ans: $H = 315.473407°$ since $\sin A = 0.906308$)
9. Compute
   $$H = \frac{H}{15}.$$
   This converts H to hours to give the hour angle.
   (Ans: $H = 21.031560^h$)
10. Convert H and δ to HMS and DMS format.
   (Ans: $H = 21^h1^m54^s$, $\delta = 8°5'3''$)

Note that no angle adjustment is needed in step 4 since declination is in the range of −90° to +90° which matches the range of the inverse sine function. However, step 8 is required since H is in the range of 0^h to 24^h (0° to 360°) but step 7 produces an angle in the range of 0° to 180°. After step 10 is complete, H can be converted to α if desired by using equation 5.5.2 in the technique already described above.

A similar procedure is required to convert from the equatorial to the horizon coordinate system. Suppose a star is located at $\delta = -0°30'30''$, $H = 16^h29^m45^s$. For an observer at 25°N latitude, where will the star appear in the sky?

1. Convert H to decimal format.
   (Ans: $H = 16.495833^h$)
2. Multiply step 1 by 15 to convert to degrees.
   (Ans: $H = 247.437500°$)
3. Convert δ to decimal format.
   (Ans: $\delta = -0.508333°$)
4. Compute
   $$T = \sin\delta * \sin\phi + \cos\delta * \cos\phi * \cos H.$$
   (Ans: $T = -0.351477$)
5. $h = \sin^{-1}T$.
   (Ans: $h = -20.577708°$)
6. $T_1 = \sin\delta - \sin\phi * \sin h$.
   (Ans: $T_1 = 0.139669$)
7. Compute
   $$T_2 = \frac{T_1}{\cos\phi * \cos h}.$$
   (Ans: $T_2 = 0.164610$)
8. $A = \cos^{-1}T_2$.
   (Ans: $A = 80.525408°$)
9. If $\sin H > 0$, $A = 360° - A$.
   (Ans: $A = 80.525408°$ since $\sin H = -0.923462$)
10. Convert A and h to DMS format.
   (Ans: $A = 80°31'31''$, $h = -20°34'40''$)

Note that since the altitude is negative in this example, the observer will not be able to see the star. Notice also that no angle adjustment is needed in step 5. Why? Step 9 calculates an azimuth adjustment. Why?

5.7 Ecliptic Coordinate System

Equatorial and horizon coordinates are not terribly convenient for calculations involving the location of objects within the solar system. Ecliptic coordinates are used instead for computing positions and then converted to equatorial or horizon coordinates. Because objects within the solar system are constantly moving, their ecliptic coordinates (and consequently their equatorial coordinates) change appreciably over months or even weeks.

Figure 5.10 illustrates the relationship between the ecliptic plane and a planet's true orbit. The ecliptic plane plays the role of "ecliptic equator" so that ecliptic latitudes are measured with respect to the ecliptic plane. Latitudes above the ecliptic plane are positive angles while latitudes below the ecliptic plane are negative angles. Thus, ecliptic latitudes are in the range $-90°$ to $+90°$. Objects, such as the Sun, which lie entirely within the ecliptic plane always have an ecliptic latitude of $0°$. Ecliptic latitude is designated by the symbol β.

To determine ecliptic longitude, the First Point of Aries again plays a role similar to Greenwich, England. Ecliptic longitudes range from $0°$ to $360°$. The plane of the ecliptic and the plane containing the celestial equator are at an angle to each other. This angle of inclination is called the obliquity of the ecliptic and is about $23°26'$. The intersection of the ecliptic plane and the celestial equator forms a line which is a diameter of the celestial sphere. This diameter intersects the celestial sphere in two points, one on either side of the celestial sphere. Ecliptic longitude is designated by the symbol λ.

Converting between ecliptic and equatorial coordinate systems requires knowing the obliquity of the ecliptic, denoted by ε. An average value is $23°26'$ but for better accuracy the angle should be computed since it varies slowly with time. This will be demonstrated momentarily.

Due to an effect called precession of the equinoxes (see below), right ascension and declination for a given star change very slowly with time. Therefore, when a right ascension and declination are given, it is customary to indicate upon what instant of time the measurements are based. This instant of time is called a standard epoch. Many star atlases and star catalogues give equatorial coordinates for epoch 1950.0 while many newer ones are for epoch 2000.0. Unless high accuracy is needed, the epoch can often be ignored. However, when computing the obliquity of the ecliptic, the calculations should be made for the instant of time corresponding to the standard epoch. For example, if the right ascension and declination are to be based upon standard epoch 2000, then the obliquity of the ecliptic should be calculated for the standard epoch 2000. Be sure to watch out and account for this in the chapters ahead. Most will assume a standard epoch of 1980.0.

The equation for computing the obliquity of the ecliptic is

$$\varepsilon = \varepsilon_0 - \frac{46.845T + 0.0059T^2 - 0.00181T^3}{3,600} \qquad (5.7.1)$$

where $\varepsilon_0 = 23.452294$ is the obliquity of the ecliptic at epoch 1900.0 and T is the number of Julian centuries since 1900.0. Using this equation, four steps are required to compute ε. Assume the standard epoch is 1950.0.

1. Compute JD, the Julian Date, for 1/0.0/Epoch. (Ans: $JD = 2,433,281.5$)
2. Compute T, the number of Julian centuries since 1/0.5/1900.
$$T = \frac{JD - 2,415,020.0}{36,525}.$$
(Ans: $T = 0.499973$ centuries)
3. $D_e = 46.845 * T + 0.0059 * T^2 - 0.00181 * T^3$. (Note: Roundoff error can be reduced if this is calculated in the equivalent form
$$D_e = T * [46.845 + T * (0.0059 - 0.00181 * T)]$$
but this enhancement will be ignored for the sake of clarity.
(Ans: $D_e = 23.422466$)
4. Finally,
$$\varepsilon = \varepsilon_0 - \frac{D_e}{3600}.$$
This is the desired obliquity of the ecliptic.
(Ans: $\varepsilon_{1950.0} = 23.445788°$)

ε could be converted to the DMS format in step 4, but it is left in the decimal format for ease of use in other procedures.

Given the obliquity of the ecliptic, the ecliptic and equatorial coordinate systems are related by the following four equations:

$$\sin\delta = \sin\beta * \cos\varepsilon + \cos\beta * \sin\varepsilon * \sin\lambda \qquad (5.7.2)$$

$$\tan\alpha = \frac{\sin\lambda * \cos\varepsilon - \tan\beta * \sin\varepsilon}{\cos\lambda} \qquad (5.7.3)$$

$$\sin\beta = \sin\delta * \cos\varepsilon - \cos\delta * \sin\varepsilon * \sin\alpha \qquad (5.7.4)$$

$$\tan\lambda = \frac{\sin\alpha * \cos\varepsilon + \tan\delta * \sin\varepsilon}{\cos\alpha}. \qquad (5.7.5)$$

Notice the similarity between these two pairs of equations. The pairs are identical when δ is interchanged with β and α is interchanged with λ . All measurements in these equations are expressed as angles so that α must be converted from HMS format to DMS format before using the equations.

Suppose the ecliptic coordinates of Jupiter are ecliptic longitude $184°36'$ and ecliptic latitude $1°12'$. Find the equatorial coordinates of Jupiter for the standard epoch 1950.0.

1. Calculate ε for the given epoch.
   (Ans: $\varepsilon_{1950.0} = 23.445788°$)
2. Convert λ to decimal format.
   (Ans: $\lambda = 184.6°$)
3. Convert β to decimal format.
   (Ans: $\beta = 1.2°$)
4. Compute
   $$\sin\delta = \sin\beta * \cos\varepsilon + \cos\beta * \sin\varepsilon * \sin\lambda.$$
   (Ans: $\sin\delta_{1950.0} = -0.012689$)
5. Take the inverse sine of step 4 to get δ.
   (Ans: $\delta_{1950.0} = -0.727068°$)
6. $y = \sin\lambda * \cos\varepsilon - \tan\beta * \sin\varepsilon$.
   (Ans: $y = -0.081912$)
7. $x = \cos\lambda$.
   (Ans: $x = -0.996779$)
8. Compute
   $$R = \tan^{-1}\left(\frac{y}{x}\right).$$
   (Ans: $R = 4.697829°$)
9. Compute an angle adjustment for R based upon y and x to put the result of $\tan^{-1}$ in the correct quadrant.
   (Ans: *Adjustment* $= 180°$)
10. $R = R +$ *Adjustment*.
    (Ans: $R = 184.697829°$)
11. Convert R to hours to give α in decimal hours.
    $$\alpha = \frac{R}{15}.$$
    (Ans: $\alpha_{1950.0} = 12.313189^{\mathrm{h}}$)
12. Convert α and δ to HMS and DMS format.
    (Ans: $\alpha_{1950.0} = 12^{\mathrm{h}}18^{\mathrm{m}}47^{\mathrm{s}}$, $\delta_{1950.0} = -0°43'37''$)

Let us illustrate converting right ascension and declination to ecliptic longitude and ecliptic latitude by doing the previous problem in reverse. The right ascension is $12^{\mathrm{h}}18^{\mathrm{m}}47^{\mathrm{s}}$, declination is $-0°43'37''$, and the standard epoch is 1950.0.

1. Calculate ε for the given epoch.
   (Ans: $\varepsilon_{1950.0} = 23.445788°$)
2. Convert α to decimal format.
   (Ans: $\alpha_{1950.0} = 12.313056^{\mathrm{h}}$)
3. Convert δ to decimal format.
   (Ans: $\delta_{1950.0} = -0.726944°$)
4. Convert α from the previous step to degrees.
   $$\alpha = 15 * \alpha.$$
   (Ans: $\alpha = 184.695833°$)
5. Compute
   $$\sin\beta = \sin\delta * \cos\varepsilon - \cos\delta * \sin\varepsilon * \sin\alpha.$$
   (Ans: $\sin\beta_{1950.0} = 0.020931$)
6. Take the inverse sine of step 5 to get the ecliptic latitude.
   (Ans: $\beta_{1950.0} = 1.199332°$)
7. $y = \sin\alpha * \cos\varepsilon + \tan\delta * \sin\varepsilon$.
   (Ans: $y = -0.080156$)
8. $x = \cos\alpha$.
   (Ans: $x = -0.996643$)
9. Compute
   $$R = \tan^{-1}\left(\frac{y}{x}\right).$$
   (Ans: $R = 4.598153°$)
10. Compute a quadrant adjustment for R based upon y and x.
    (Ans: *Adjustment* $= 180°$)
11. $\lambda = R +$ *Adjustment*.
    (Ans: $\lambda_{1950.0} = 184.598153°$)
12. Convert the ecliptic longitude and the ecliptic latitude to the DMS format.
    (Ans: $\lambda_{1950.0} = 184°35'53''$, $\beta_{1950.0} = 1°11'58''$)

The discrepancy between the result for this example and the previous example is due to roundoff errors.

5.8 Galactic Coordinate System

The last coordinate system to be considered is the galactic coordinate system. In this coordinate system, the plane containing the Sun and the center of our galaxy, the Milky Way Galaxy, intersects the celestial sphere to form a "galactic equator." The coordinates for the center of our galaxy is about right ascension $17^{\mathrm{h}}42^{\mathrm{m}}$ and declination $-28°45'$. Galactic latitudes, denoted by the symbol l, are in the range of $-90°$ to $+90°$ with positive angles being North of the galactic plane and negative angles being South of the galactic plane. Extension of a line connecting the Sun and the center of the galaxy until it intersects the celestial sphere will form two points on the celestial sphere. The intersection point farthest away from the Sun is chosen as the reference point for measuring galactic longitudes, denoted by the symbol b. Earth longitudes and ecliptic longitudes are measured in a clockwise manner but the galactic longitude is measured counterclockwise so as to increase from 0° to 360° in the same direction as increasing right ascension.

For epoch 1950.0, the location of the Galactic North Pole is right ascension 12^h49^m, declination 27°24′. (Note: It will be more convenient in the algorithm below to convert α and δ to decimal degrees. Thus, $\alpha = 192.25°$ and $\delta = 27.4°$.) The ascending node of the galactic plane is 33°. Strictly speaking, these three angles should be adjusted to refer to the same standard epoch as that of the desired right ascension and declination. Since galactic coordinates will not be used beyond this chapter, this adjustment will be omitted. The next section will demonstrate how to compute precession corrections to replace the equatorial coordinates given here for the Galactic North Pole. Once the corrected coordinates are known, replace the right ascension (192.25°) and declination (27.4°) in the procedure below with the new corrected equatorial coordinates. This will refer the galactic coordinates to the new standard epoch.

There is an easier way, however. The procedure below to convert galactic to equatorial coordinates produces epoch 1950.0 coordinates. Once calculated, these coordinates can be adjusted for precession. In converting equatorial to galactic coordinates, if the equatorial coordinates are not epoch 1950.0, adjust them to that epoch first. Then convert the new epoch 1950.0 equatorial coordinates to galactic coordinates.

The equations relating galactic and equatorial coordinates are

$$\begin{aligned}
\sin\delta &= \cos l * \cos\delta_0 * \sin(b - N_0) + \sin l * \sin\delta_0 \\
\tan\alpha &= \frac{\cos l * \cos(b - N_0)}{\sin l * \cos\delta_0 - \cos l * \sin\delta_0 * \sin(b - N_0)} \\
\sin l &= \cos\delta * \cos\delta_0 * \cos(\alpha - \alpha_0) + \sin\delta * \sin\delta_0 \\
\tan b &= \frac{\sin\delta - \sin l * \sin\delta_0}{\cos\delta * \sin(\alpha - \alpha_0) * \cos\delta_0}.
\end{aligned}$$

where $\delta_0 = 27.4°$ and $\alpha_0 = 192.25°$ are the epoch 1950.0 equatorial coordinates of the Galactic North Pole, and $N_0 = 33°$ is the ascending node of the galactic plane.

Assuming epoch 1950.0, convert galactic longitude 180°, galactic latitude 55°20′ to equatorial coordinates.

1. Convert b to decimal format.
   (Ans: $b_{1950.0} = 180.0°$)
2. Convert l to decimal format.
   (Ans: $l_{1950.0} = 55.333333°$)
3. Compute
   $$\sin\delta = \cos l * \cos\delta_0 * \sin(b - N_0) + \sin l * \sin\delta_0.$$
   where the constants δ_0 and N_0 are the right ascension and ascending node of the Galactic North Pole respectively for epoch 1950.0. ($\delta_0 = 27.4°$, $N_0 = 33°$)
   (Ans: $\sin\delta_{1950.0} = 0.653540$)
4. Take the inverse sine of step 3 to get the declination.
   (Ans: $\delta_{1950.0} = 40.809051°$)
5. $y = \cos l * \cos(b - N_0)$.
   (Ans: $y = -0.477037$)
6. $x = \sin l * \cos\delta_0 - \cos l * \sin\delta_0 * \sin(b - N_0)$.
   (Ans: $x = 0.587640$)
7. Compute
   $$R = \tan^{-1}\left(\frac{y}{x}\right).$$
   (Ans: $R = -39.069122°$)
8. Calculate an adjustment for R based upon y and x.
   (Ans: *Adjustment* = 360°)
9. $\alpha = R + \textit{Adjustment} + \alpha_0$ where the constant $\alpha_0 = 192.25°$ is the declination of the Galactic North Pole for epoch 1950.0.
   (Ans: $\alpha_{1950.0} = 513.180878°$)
10. If $\alpha > 360°$, subtract 360°.
    (Ans: $\alpha_{1950.0} = 153.180878°$)
11. Convert α to hours.
    $$\alpha = \frac{\alpha}{15}.$$
    (Ans: $\alpha_{1950.0} = 10.212059^h$)
12. Convert α and δ to HMS and DMS format.
    (Ans: $\alpha_{1950.0} = 10^h12^m43^s$, $\delta_{1950.0} = 40°48'33''$)

Again assuming epoch 1950.0, convert right ascension $10^h12^m43^s$, declination 40°48′33″ to their equivalent galactic coordinates.

1. Convert δ to decimal format.
   (Ans: $\delta_{1950.0} = 40.809167°$)
2. Convert α to decimal format.
   (Ans: $\alpha_{1950.0} = 10.211944^h$)
3. Convert α to degrees.
   $$\alpha = 15 * \alpha.$$
   (Ans: $\alpha_{1950.0} = 153.179167°$)

4. Compute

$$\sin l = \cos\delta * \cos\delta_0 * \cos(\alpha - \alpha_0) + \sin\delta * \sin\delta_0.$$

(Ans: $\sin l_{1950.0} = 0.822462$)
5. Take the inverse sine to obtain the galactic latitude.
(Ans: $l_{1950.0} = 55.332050°$)
6. $y = \sin\delta - \sin l * \sin\delta_0$.
(Ans: $y = 0.275045$)
7. $x = \cos\delta * \sin(\alpha - \alpha_0) * \cos\delta_0$.
(Ans: $x = -0.423535$)
8. Compute

$$T = \tan^{-1}\left(\frac{y}{x}\right).$$

(Ans: $T = -32.999769°$)
9. Compute a quadrant adjustment for T based upon y and x.
(Ans: $Adjustment = 180°$)
10. $b = T + Adjustment + N_0$.
(Ans: $b_{1950.0} = 180.000231°$)
11. If $b > 360°$, subtract 360°.
(Ans: $b_{1950.0} = 180.000231°$)
12. Convert b and l to the DMS format.
(Ans: $b_{1950.0} = 180°0'1''$, $l_{1950.0} = 55°19'55''$)

5.9 Precession and Other Corrections

Several factors can affect the apparent location of objects within the coordinate systems presented. Precession, refraction, parallax, and nutation will be briefly described in this section but a correction algorithm will be presented only for precession.

Earth is surrounded by an atmosphere. Light reaching us from objects outside the Earth must pass through the atmosphere before we can see it. Light rays are bent when they enter the atmosphere so that stars and other objects do not appear to be quite where they are calculated to be. This effect is called atmospheric refraction. The extent to which it distorts the location of an object depends upon air temperature, air pressure, and the object's altitude. Objects at the zenith are not refracted at all while those on the horizon are refracted a great deal. Atmospheric refraction can be as much as 34′ of arc for objects at the horizon.

Parallax is an effect due to observers being in different locations on the Earth. This effect is most noticeable to the naked eye when looking at the Moon although the same problem occurs to a lesser extent when viewing planets and is virtually unmeasurable for stars. Stellar parallax, measured from the Earth's orbit, is an analogous effect observable from different places along the Earth's orbit. Figure 5.13 shows two observers at location A and B who are looking at the Moon. S is a distant star. Since S is so far away relative to the distance between observer A and observer B, both observers will agree upon its location. However, observers A and B will not agree upon the location of the Moon with respect to the star S. The angle MAS measured by observer A is not the same as $\angle MBS$ measured by observer B. Parallax distortion can amount to one degree of arc for viewing the Moon but is only a few seconds of arc when viewing planets.

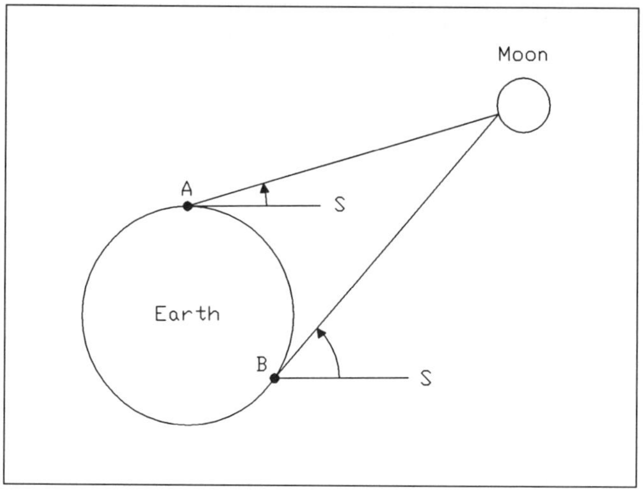

Figure 5.13: *Parallax is the result of viewing an object from different locations.*

The last two effects to be described are nutation and precession. Nutation is a periodic wobbling of the Earth's axis which may affect an object's apparent location by a few seconds of arc. Precession is a motion of the Earth's axis very similar to the motion of a rapidly spinning top. When a top spins, its axis of rotation actually moves in a relatively small circle. This motion of the rotation axis is called precession, or more precisely, precession of the equinoxes. In the case of the Earth, the combined gravitational pull of the Sun and the Moon cause precession. The North Pole of the Earth rotates around the pole of the ecliptic once in about every 25,800 years. Precession must be accounted for to obtain high precision. It is presented here only as an example of how equatorial coordinates can be adjusted for various effects.

Calculating precession corrections requires knowing the equatorial coordinates of interest, the reference epoch for the coordinates, and the new standard epoch. The two equations required to compute the precession corrections for α and δ are

$$\Delta\alpha = (M + N_t * \sin\alpha_{old} * \tan\delta_{old}) * D \quad (5.9.1)$$
$$\Delta\delta = N_d * \cos\alpha_{old} * D \quad (5.9.2)$$

where D is the difference from the new epoch to the old epoch, and N_d, N_t, and M are corrections calculated from interpolation polynomials that are a function of T, the number of Julian centuries from the desired new epoch to the standard epoch 1900.0. The interpolating polynomials are

$$M = 3.07234^{s} + 0.00186^{s} * T \quad (5.9.3)$$
$$N_d = 20.0468'' - 0.0085'' * T \quad (5.9.4)$$
$$N_t = \frac{N_d}{15}. \quad (5.9.5)$$

Suppose that the epoch 1950.0 equatorial coordinates of an object are right ascension $18^{h}30^{m}12^{s}$, declination $-85°16'30''$. Find the new equatorial coordinates for epoch 1980.0.

1. Convert α to decimal format.
   (Ans: $\alpha_{1950.0} = 18.503333^{h}$)
2. Multiply α by 15 to convert it to degrees.
   (Ans: $\alpha_{1950.0} = 277.55°$)
3. Convert δ to decimal format.
   (Ans: $\delta_{1950.0} = -85.275°$)
4. Let E_t be the epoch to convert to and compute
   $$T = \frac{E_t - 1900}{100}.$$
   (Ans: $T = 0.8$)
5. Compute $M = 3.07234^{s} + 0.00186^{s} * T$ where M is in seconds of time.
   (Ans: $M = 3.073828^{s}$ of time)
6. Compute $N_d = 20.0468'' - 0.0085'' * T$ where N_d is in seconds of arc.
   (Ans: $N_d = 20.04''$)
7. Compute
   $$N_t = \frac{N_d}{15}.$$
   This converts arcseconds to seconds of time.
   (Ans: $N_t = 1.336^{s}$)
8. Let E_f be the epoch to convert from and compute $D = E_t - E_f$.
   (Ans: $D = 30$)
9. $\Delta\alpha = (M + N_t * \sin\alpha * \tan\delta) * D$.
   This is an adjustment factor for the right ascension in seconds of time.
   (Ans: $\Delta\alpha = 572.923219^{s}$ of time)
10. $\Delta\delta = N_d * \cos\alpha * D$.
    This is a correction for the declination in arcseconds.
    (Ans: $\Delta\delta = 78.992854''$ of arc)
11. Divide $\Delta\alpha$ by 3600 to convert it to hours and divide $\Delta\delta$ by 3600 to convert it to degrees.
    (Ans: $\Delta\alpha = 0.159145^{h}$, $\Delta\delta = 0.021942°$)
12. Compute
    $$\alpha_{1980.0} = \Delta\alpha + \alpha_{1950.0}$$
    and
    $$\delta_{1980.0} = \Delta\delta + \delta_{1950.0}$$
    to obtain the corrected equatorial coordinate for the new epoch.
    (Ans: $\alpha_{1980.0} = 18.662479^{h}$, $\delta_{1980.0} = -85.253058°$)
13. Convert α and δ to HMS and DMS format.
    (Ans: $\alpha_{1980.0} = 18^{h}39^{m}45^{s}$, $\delta_{1980.0} = -85°15'11''$)

5.10 About the Program

The program COORD implements the algorithms discussed in this chapter. Since many implementations of BASIC do not include inverse trigonometric functions, COORD contains functions which define the inverse sine, inverse cosine, and inverse tangent. Moreover, most computer languages assume that the angle is to be expressed in radians rather than degrees. Functions are also included in the program which define sine, cosine, and tangent to assume the angles are expressed in degrees. The inverse trigonometric functions are written to return angles in degrees rather than radians. The programs for later chapters will include these function definitions.

Location angles, such as δ are required by the program. Location angles may be entered in decimal or DMS format. To enter an angle in decimal format, the decimal value is typed in when the degrees are asked for and a value of 0 is entered in response to the request for arcminutes and arcseconds. When entering a DMS format value, be sure to enter a negative sign if the angle is negative but greater than -1. That is, the angle $-0°40'30''$ should be entered as -0 for degrees, 40 for minutes, and 30 for seconds. Along with watching out for a negative sign for negative angles, enter a direction of N or S for latitudes and E or W for longitudes.

5.11 Sample Problems

The example problems which follow should be thoroughly understood before proceeding to the remaining chapters. Each chapter which follows will perform frequent coordinate system conversions to locate astronomical objects.

1. A certain star has right ascension $12^h32^m6^s$. An observer is at 40°E longitude. The date is 1/5/1980 and GMT is 12^h. What is the hour angle for this star?
(Ans: $H = 9^h4^m54^s$)
2. A star with hour angle $15^h30^m15^s$ is observed from longitude 64°W. The date is 6/5/1976 and GMT is 14^h. What is the star's right ascension?
(Ans: $\alpha = 11^h10^m13^s$)
3. A celestial body is seen in the sky at altitude 10°0′0″, azimuth 200°10′20″. The observer's latitude is 35.6° N. What is the hour angle and declination of the object?
(Ans: $H = 1^h46^m15^s$, $\delta = -40°34'58''$)
4. The hour angle for a certain star is $7^h0^m0^s$. Its declination is 49°54′20″. If the observer is at latitude 80°S, where will this star appear in the sky?
(Ans: $h = -51°28'21''$, $A = 267°7'4''$. The star is below observer's horizon.)
5. The ecliptic location for Mercury is computed to be ecliptic longitude 120°30′30″, ecliptic latitude 0°0′0″. What are the equatorial coordinates for Mercury assuming epoch 1950.0?
(Ans: $\alpha_{1950.0} = 8^h10^m51^s$, $\delta_{1950.0} = 20°2'50''$)
6. For epoch 2000.0, assume that a celestial body is at right ascension $11^h10^m13^s$, declination 30°5′40″. What are it's ecliptic coordinates?
(Ans: $\lambda_{2000.0} = 156°19'9''$, $\beta_{2000.0} = 22°41'54''$)
7. A star within the Milky Way Galaxy is at galactic latitude 30°25′40″, galactic longitude 120°0′0″. What are it's equatorial coordinates?
(Ans: $\alpha = 15^h29^m53^s$, $\delta = 85^h59^m33^s$)
8. If a star's location is right ascension $11^h10^m13^s$, declination 30°5′40″, what are it's galactic coordinates?
(Ans: $b = 200°0'15''$, $l = 68°13'25''$)
9. If the epoch 1950.0 equatorial coordinates are right ascension $12^h32^m6^s$, declination 30°5′40″, calculate the epoch 2000.0 coordinates by correcting for precession.
(Ans: $\alpha_{2000.0} = 12^h34^m34^s$, $\delta_{2000.0} = 29°49'8''$)
10. If the epoch 1985.5 equatorial coordinates are right ascension $11^h10^m13^s$, declination 30°5′40″, calculate the epoch 1950.0 coordinates by correcting for precession.
(Ans: $\alpha_{1950.0} = 11^h8^m18^s$, $\delta_{1950.0} = 30°17'15''$)

Chapter 6

Stars in the Night Sky

At this point, the reader is most likely confused. Chapter 4 discussed time and presented a maze of calculations for converting from one time system to another. But *why* is all that necessary? Different calendar systems were described and techniques presented for dealing with Julian dates. *Why* are Julian dates necessary?

Chapter 5 may have appeared to the reader to be a bit more illuminating. Different coordinate systems were presented with algorithms for converting between them. The horizon and equatorial coordinate systems in particular may have appeared very promising and left the reader feeling that he was on the brink of understanding ... something. But still, *how* does time fit in with the coordinate system conversions?

The basic question is this: For a given date, time, longitude, and latitude, where will the Sun, Moon, planets, and stars appear in the night sky? All the pieces seem to be here, but how do they fit together? The present chapter will demonstrate *how* and *why* the pieces fit together to locate a specific star. In fact, this chapter includes a program that will produce a star chart which shows the night sky for a given time, date, longitude, and latitude! The location of the Sun, Moon, and planets must wait for a little while longer.

To determine where a star is in the sky, its equatorial coordinates must be known. These coordinates can be obtained from a star catalogue or a star atlas. Simply take the equatorial coordinates given, use them as input to the STARS program for this chapter, and as if by magic, out comes the horizon coordinates which tell where to point the telescope.

The material in this chapter is not limited to just locating stars. Any object whose equatorial coordinates are known will work as well. For example, one could consult one of the popular astronomy periodicals to get the equatorial coordinates for a planet and then use the program STARS to locate the planet. However, the equatorial coordinates of objects within the solar system change very rapidly with time. The trick, then, is to be able to calculate a planet's equatorial coordinates. The remaining chapters will show how to calculate the equatorial coordinates of solar system objects and then use the results of this chapter to locate them in the sky.

Some license is taken in the discussion which follows by omitting or "glossing over" some technical details that might be more confusing than illuminating. This is done to promote conceptual understanding by providing a framework in which all the pieces will fit. May the purists be forgiving for this simplified treatment. Nonetheless, clean off the table and get some glue ready. The pieces are all assembled. Now it is time to glue them together to form a picture of mathematical beauty and elegance. The result should be a much better understanding, albeit very basic, of how the universe seems to work. Chapters 7, 8, and 9 will serve to put some color into this black and white picture.

6.1 Locating a Star

To locate a star in horizon coordinates, it is necessary to account for three things: date, time of day, and observer's location. Julian day numbers provide a convenient (well, at least manageable) method for accounting for dates. Julian day numbers are especially useful for computing elapsed days and many of the calculations, such as converting GMT to GST, do precisely that. But why are elapsed days important?

To answer the question just raised, assume that the Earth and Sun are stationary and the stars, embedded in the celestial sphere, move around the Earth. If a snapshot is taken of the position of the stars relative to the Earth, then in exactly one year (sidereal year) the stars will return to precisely the same location in the sky. That is, the stars will have moved through 360° along the celestial sphere. If only half a sidereal year has elapsed, then the stars will only have moved 180° along the celestial sphere. Since there are 366.2596 sidereal days in a sidereal year, the stars move by 1.0174° per sidereal day. Therefore, if N days have elapsed, then the stars will have moved by $1.0174 * N$ degrees. The date can be "canceled out" and accounted for by using Julian day numbers to calculate elapsed days. This is precisely what steps 4–7 in converting GMT to GST are all about.

Of course it is well known that the Earth rotates on

its axis. Therefore, the time of day affects the apparent motion of the stars in the sky. Actually, time of day has three effects which must be considered:

1. Time Zone
2. Sidereal Time
3. Time of Day.

First, time zone differences are treated by converting LMT to GMT. Secondly, our wristwatches measure mean solar time while celestial motion is measured by "star" time. Converting GMT to GST accounts for this difference in time units. Thirdly, converting GST to LST accounts for how much the Earth has rotated during the fractional part of a day represented by the LMT.

Recall that the GST to LST conversion uses the observer's longitude to arrive at the LST. An observer's longitude is just another way of expressing how far the Earth has rotated from 0^h until the calculated GMT. But the important thing to realize is that the time conversion process from LMT to LST is to "synchronize clocks" so that "star" time is being used.

Up to this point calculations have been made to "cancel out" effects of the date, time of day, and longitude. Only two small problems remain. First, the equatorial coordinates are independent of time of day but they somehow need to be converted to a format which is time (sidereal) dependent. Secondly, an observer's latitude must be considered. The first problem is solved by converting the right ascension to the hour angle and using the hour angle in the calculation of altitude and azimuth. Note that the hour angle is indeed time varying. Adjustments for latitude are made in steps 4–7 of the procedure for converting equatorial coordinates to horizon coordinates.

Table 6.1 more or less summarizes the steps required and why they are needed for locating a star given an observer's location, LMT, and date. The effects of precession, nutation, refraction, and parallax can also be included if desired. Apply the corrections to the equatorial coordinates before beginning the process outlined in Table 6.1. Chapters 7, 8, and 9 will fit into the process very easily. These chapters will demonstrate how to calculate the equatorial coordinates for the Sun, Moon, and planets which can then be transformed into horizon coordinates as shown here.

Once broken down into a series of major blocks, the process is not so bewildering after all. But before working out an example, a word of warning is in order. Converting GMT to GST requires a computation involving the date as does the Julian date calculation. However, it is possible that the date will not be the same for the LMT and GMT times. When adding the zone adjustment to LMT, the resulting GMT might be greater than 24^h or less than 0^h. If the GMT is greater than 24^h, then 24^h is subtracted from the GMT to put it into the proper range. When this occurs, the resulting GMT is actually for the next day so the date must be adjusted accordingly. Likewise, if the GMT is negative, 24^h is added to the GMT with the result that the GMT is actually for the previous day and the date must be adjusted accordingly. (Watch out for this adjustment in the subroutine converting LMT to GMT and the subroutine for GMT to LMT in the program STARS.) As a general rule, time conversions should be done first (as shown in Table 6.1) so that the GMT date is used for the Julian date and GST calculations.

Calculation	Effect
LMT to GMT	Account for time zone
GMT to GST	"Synchronize" clocks
GST to LST	Account for longitude
Date to JD	Elapsed days to account for date
RA to HA	Make equatorial coordinates time varying
Eq. to Horizon	Adjust for latitude

Table 6.1 *Converting Equatorial to Horizon Coordinates requires these fundamental steps.*

Now let us proceed with an example and compute the horizon coordinates for Saturn. Assume an observer at longitude 78°W, latitude 38°N is in the Eastern Standard Time Zone. He looks in *Sky & Telescope* to find that on February 14, 1985 Saturn will be at about right ascension $15^h44^m6^s$, declination $-17°32'0''$. At $21^h30^m0^s$ LMT, where will Saturn appear for the observer? Assume that the observer is not on Daylight Saving Time and ignore precession.

1. Convert LMT to decimal format.
   (Ans: $LMT = 21.5^h$)
2. Convert LMT to GMT.
   (Ans: $GMT = 2.5^h$, next day so use the date 2/15/1985)
3. Convert GMT to GST.
   (Ans: $GST = 12.169870^h$)
4. Convert GST to LST.
   (Ans: $LST = 6.969870^h$)
5. Convert α (right ascension) to H (hour angle).
   (Ans: $H = 15.234870^h$)
6. Convert equatorial to horizon coordinates.
   (Ans: $h = -43.089617°$, $A = 78.033009°$)

7. Convert h and A to DMS format.
(Ans: $h = -43°5'23''$, $A = 78°1'59''$)

Notice that Saturn will not be visible at this time since it's altitude is below the observer's horizon. Also notice that the date must be adjusted because of step 2.

Sometimes it is desirable to reverse the process above and calculate equatorial coordinates for a given set of horizon coordinates. At $21^h45^m0^s$ LMT, the observer from the previous example notices a bright star at approximately altitude 54°44′, azimuth 215°6′. What are the object's equatorial coordinates?

1. Convert LMT to decimal format.
(Ans: $LMT = 21.75^h$)
2. Convert LMT to GMT.
(Ans: $GMT = 2.75^h$, next day so use the date 2/15/1985)
3. Convert GMT to GST.
(Ans: $GST = 12.420555^h$)
4. Convert GST to LST.
(Ans: $LST = 7.220555^h$)
5. Convert horizon to equatorial coordinates.
(Ans: $H = 1.304263^h$, $\delta = 7.494255°$)
6. Convert H to α.
(Ans: $\alpha = 5.916292^h$)
7. Convert α and δ to HMS and DMS format.
(Ans: $RA = 5^h54^m59^s$, $\delta = 7°29'39''$)

A star chart shows that the star Betelgeuse in the constellation Orion is very near this location.

6.2 Star Rising and Setting Times

The preceding section has one small problem with it. For a given LMT, the horizon coordinates for a star can be calculated easily enough, but there is no guarantee that the star will be visible. There are two reasons why a star might not be visible. First of all, a star might not be visible because the observer's location is such that the star does not appear above the horizon. For instance, an observer at the South Pole will never be able to see Polaris, the Pole Star, because it is never visible in the Southern Hemisphere. Saturn in the example above is not visible to our observer for the given LMT because it is below the horizon for our observer's location. Secondly, a star might not be visible because it rises and sets during daylight hours. Unless an eclipse occurs, sunlight will prevent one from being able to see a star even if it is above the horizon.

This section will demonstrate how to calculate the rising and setting times for a given set of equatorial coordinates. Once it is known when a star will be above the horizon, an observer can choose a convenient viewing time (assuming it will appear at night!) and then calculate where the star will appear. More will be said in just a moment about stars which do not rise or set.

Determining the star rise and set times is accomplished by determining the LST rising and setting times and then converting the LST times to LMT times. Using the subscript r to denote rising time and the subscript s to denote setting time, the required LST equations are

$$LST_r = 24^h + \alpha - \frac{\cos^{-1}(-\tan\phi * \tan\delta)}{15}$$
$$LST_s = 24^h + \frac{\cos^{-1}(-\tan\phi * \tan\delta)}{15}$$

where ϕ is the observer's latitude and α and δ are the star's equatorial coordinates.

To demonstrate the process, assume that the observer is the same as above. He is located at longitude 78°W, latitude 38°N. What are the rising and setting times for Betelgeuse ($\alpha = 5^h55^m$, $\delta = 7°30'$) on February 14, 1985? In the calculations below, remember that the subscript r refers to rising times and the subscript s refers to setting times.

1. Convert α to decimal format.
(Ans: $\alpha = 5.916667^h$)
2. Convert δ to decimal format.
(Ans: $\delta = 7.5°$)
3. Compute the value
$$A_r = \frac{\sin\delta}{\cos\phi}.$$
(Ans: $A_r = 0.165640$)
4. If $|A_r| > 1$, the star doesn't rise or set.
(Ans: Star does rise and set since $A_r = 0.165640 < 1$)
5. $R = \cos^{-1} A_r$.
(Ans: $R = 80.465577°$)
6. $S = 360 - R$.
(Ans: $S = 279.534423°$)
7. $H_1 = \tan\phi * \tan\delta$.
(Ans: $H_1 = 0.102858$)
8. If $|H_1| > 1$ then the star doesn't rise or set.
(Ans: Star does rise and set since $H_1 = 0.102858 < 1$)
9. Calculate
$$H_2 = \frac{\cos^{-1}(-H_1)}{15}.$$
(Ans: $H_2 = 6.393586^h$)

10. $LST_r = 24^h + \alpha - H_2$. This is the LST for when the star rises above the observer's horizon. (Ans: $LST_r = 23.523081^h$)
11. If $LST_r > 24^h$, subtract 24^h. (Ans: $LST_r = 23.523081^h$)
12. $LST_s = \alpha + H_2$. This is the LST for when the star sets below the observer's horizon. (Ans: $LST_s = 12.310252^h$)
13. If $LST_s > 24^h$, then subtract 24^h. (Ans: $LST_s = 12.310252^h$)
14. Convert LST_r and LST_s to LMT_r and LMT_s. These are the star rise and set LMT times. (Ans: $LMT_r = 14.073552^h$, $GMT_r = 19.073552^h$, $LMT_s = 2.891334^h$, $GMT_s = 7.891334^h$)
15. Convert LMT_r and LMT_s to HMS format. (Ans: $LMT_r = 14^h4^m25^s$, $LMT_s = 2^h53^m29^s$)

Steps 4 and 8 are necessary because the inverse cosine function is not defined for values less than -1 or values greater than $+1$. For this algorithm, the physical interpretation is that the star never rises or sets for the observer when A_r or H_1 is out of range. This does not necessarily mean that the star is not visible. Polaris, for instance, is always above the horizon for observers in the Northern Hemisphere. The algorithm will indicate that this star never rises or sets for our observer, but it will certainly be visible.

At first glance, the values of R and S in the algorithm above appear to be needless calculations. However, if the star does rise and set, then R is the azimuth when the star rises and S is the azimuth when the star sets. Hence, in the example of Betelgeuse above, the star will rise at $14^h4^m25^s$ LMT and will appear in the sky at azimuth $80°27'56''$ ($80.465577°$), altitude $0°0'0''$. Moreover, Betelgeuse will set at $2^h53^m29^s$ LMT at azimuth $279°32'4''$ ($279.534423°$), altitude $0°0'0''$.

It should be obvious from the calculation of A_r that the algorithm will not work for polar observers (Latitude 90°N or 90°S). This is because $\cos 90° = 0$ so that the computation of A_r becomes undefined. Also note that this algorithm will "blow up" for latitudes close to the Poles because as latitude approaches $-90°$ or $+90°$, the result of the division to produce A_r becomes increasingly large.

Notice in the example above that Betelgeuse will rise above the observer's horizon during daylight hours. Due to sunlight, Betelgeuse will not be visible until the Sun goes down even though it is above the observer's horizon.

The algorithm gives the times when the altitude for a star will be 0°. The effect of atmospheric refraction and other factors are not taken into account so that the apparent rise and set times will vary from what is calculated. Moreover, hills and trees can block a view of the horizon.

6.3 Star Chart

With the procedures outlined in this chapter, it is a relatively simple matter to produce star charts for a particular observer. All that is required is to take the equatorial coordinates of several stars, convert them to horizon coordinates, and plot those which are above the horizon. If BLDSTARS has already been used to enter a set of stars, then STARS includes a subroutine which can be run to produce a star chart for these stars. (The data file `STARS.DAT` is included which contains several constellations.)

Plotting the stars is a problem equivalent to making a map of the Earth's surface. Both involve plotting points from a sphere onto a plane. Unfortunately, mapping a sphere onto a plane causes distortion. Techniques for reducing distortion are beyond the scope of this book but can be investigated in texts dealing with cartography.

One way to map points from a sphere onto a plane is to do a projection of the points into a plane. Using spherical coordinates and ignoring the z–axis is one way of doing this. Figure 6.1 shows how spherical coordinates are defined. Point P is located on a sphere of radius L. P' is the projection of point P into the xy plane. Angle A is a measure of how far around P is in the xy plane while angle B is a measure of how far down P is from the z–axis. The equations for relating spherical coordinates with x, y, and z are:

$$x = L * \sin B * \cos A \qquad (6.3.1)$$

$$y = L * \sin B * \sin A \qquad (6.3.2)$$

$$z = L * \cos B. \qquad (6.3.3)$$

Figure 6.2 shows how Figure 6.1 can be projected onto a map. The axis designations in parentheses show how the map and Figure 6.1 are related. Notice that we have chosen to align the x–axis of Figure 6.1 with the y–axis of Figure 6.2 and similarly with the y–axis from Figure 6.1 and the x–axis from Figure 6.2. This choice is made so that $\angle A$ becomes identical with the definition of azimuth. Furthermore, $\angle B$ and altitude are related by the equation

$$B = 90° - h. \qquad (6.3.4)$$

Assuming $L = 1$, the equations for Figure 6.2 become

$$x = \sin(90 - h) * \sin A = \cos h * \sin A \qquad (6.3.5)$$

$$y = \sin(90 - h) * \cos A = \cos h * \cos A. \qquad (6.3.6)$$

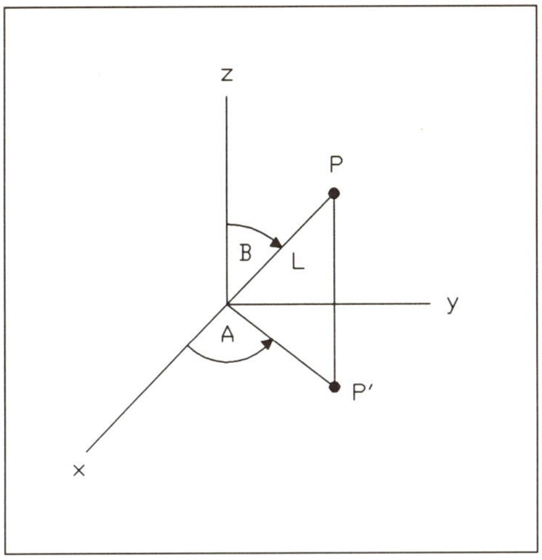

Figure 6.1: Spherical coordinate system. *The spherical coordinate system equations map the location of an object on a sphere to a plane.*

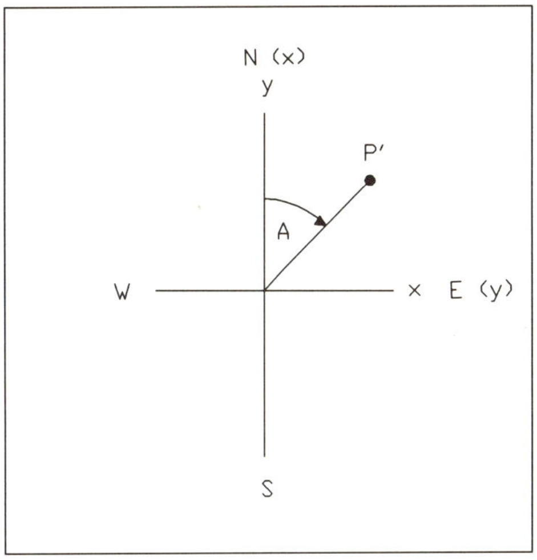

Figure 6.2: Projection onto the surface of a flat map. *This projection maps a point in 3D space to a point in 2D space.*

With the equations for Figure 6.2, producing star charts is a matter of computing the horizon coordinates, solving the mapping equations, and plotting the resulting (x, y) pairs onto a display. Appropriate scaling factors need to be considered to put each point in its proper place on the display.

The chart produced by this technique will be circular in shape. North will be at the top of the circle, South at the bottom, East to the right, and West to the left. Distortion increases as points get farther away from the center, but the charts produced serve to show the position of constellations relative to each other.

6.4 About the Program

The program STARS assumes that BLDSTARS has already been used to enter star data into a data file. If BLDSTARS has not been run yet, STARS can still be used to calculate horizon coordinates and to calculate rising and setting times but star charts can not be done. Figure 6.3 shows a picture of a star chart produced by the STARS program. Notice that Orion is clearly seen in the middle of the screen.

Several subroutines from previous programs are combined here as the basis for calculating the location of a star. In particular, the time conversions from Chapter 4 are required to convert between LMT, GMT, GST, and LST times while the coordinate system conversions from Chapter 5 are required to convert between equatorial and horizon coordinates. STARS also contains a subroutine for computing precession calculations if it is desirable to account for precession.

The subroutine which plots stars on the screen for a given observer is worth examining since it demonstrates how all the previous chapters are tied together. Stars will be plotted with North at the top, South at the bottom, East on the right, and West on the left of the screen. As each constellation is plotted, the constellation name will be displayed on the bottom left of the screen.

6.5 Sample Problems

1. An observer is located at longitude 100°W, latitude 45°N in the Pacific Standard Time Zone. The LMT is $9^h0^m0^s$ for December 1, 1984. Assuming that this is not Daylight Saving Time, calculate the horizon coordinates for a star at right ascension $6^h0^m0^s$, declination $-60°0'0''$.
(Ans: $h = -59°56'46''$, $A = 224°10'31''$)
2. An observer is located at longitude 78.3°W, latitude 38.25°N in the Eastern Standard Time Zone. At $21^h0^m0^s$ LMT on January 20, 1985, the observer locates an object at altitude $45°0'0''$, azimuth $90°0'0''$. Assuming this is not Daylight Saving Time, what are the equatorial coordinates of the object?
(Ans: $\alpha = 8^h15^m46^s$, $\delta = 25°57'41''$)
3. What are the rising and setting times for the star from problem number 1? Assume the same date and location.
(Ans: Star doesn't rise or set)
4. What are the rising and setting times for the star from problem number 2? Assume the same date and location.
(Ans: $LMT_r = 16^h57^m48^s$, $LMT_s = 7^h59^m51^s$)

Chapter 7

The Sun

Of all the celestial objects, the Sun is the most apparent and most important to life on our planet. The Sun is responsible for giving us heat and light without which life as we know it would be impossible. It is because of the Sun that we are able to see the Moon at night. Moreover, we are able to see the planets in the solar system because they also reflect light from the Sun.

In this chapter the Sun will be considered in more detail. Calculations will be presented for determining the Sun's equatorial coordinates. Of course during daylight hours on a clear day it is apparent where the Sun is in the sky. But the location of the Sun is an important consideration for determining where the planets are and certain details about the Moon. In addition to locating the Sun, calculations will be presented for determining the times of Sunrise and Sunset. The times of the equinoxes and solstices will also be calculated as well as the distance to the Sun.

Before considering some of the mathematics involving the Sun, it is appropriate to look at some facts about the Sun. First of all, the Sun is a star much like the stars we see in the night sky. Since the Sun is so much closer to the Earth than any other stars, the fact that it is a star is often overlooked. More precisely, our Sun belongs to a class of stars that are called yellow dwarfs. These appear to be a very common type of stars. Astronomers are very much interested in studying the Sun because it may explain how other stars behave.

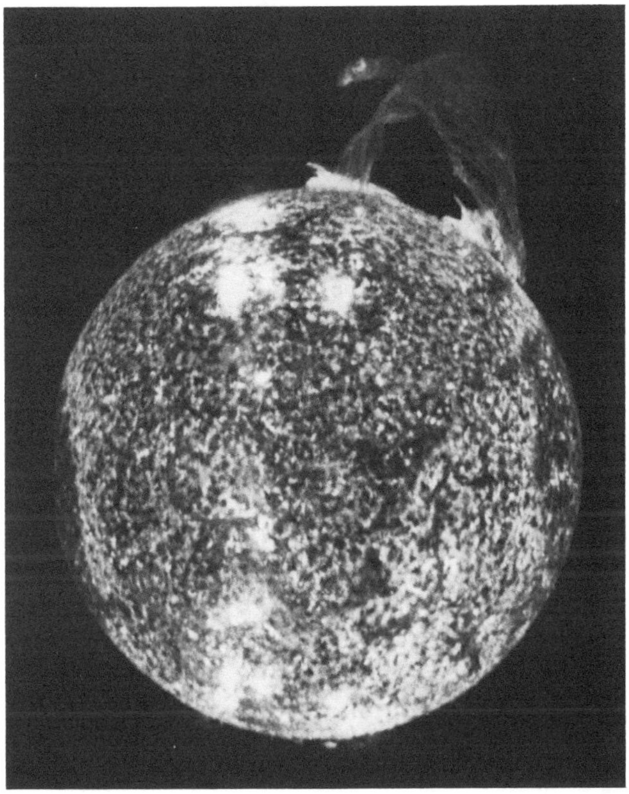

Figure 7.1: *This NASA photograph of the Sun, taken from Skylab 4, shows one of the most spectacular eruptive prominence ever recorded, spanning more than 367,000 miles. This photograph was taken in the light of ionized helium by the U.S. Naval Research Laboratory extreme ultraviolet spectroheliograph.*

By doing a spectral analysis of Sunlight, scientists have determined that the Sun is composed primarily of hydrogen and helium. Roughly 93 million miles from the Earth, the Sun is a ball that is very nearly spherical in shape as opposed to the ellipsoid shape of the Earth. Its diameter is about 1,390,600 kilometers while its mass is over 325,000 times that of the Earth. This makes the Sun by far the most massive object in the solar system. But because the Sun is made up mostly of lighter elements, its mean density is only 1.4 times that of water. The Earth by comparison has a mean density of 5.5 times that of water.

The composition of the Sun contains the key to why it gives off heat and why it shines. Our Sun is like a gigantic nuclear furnace with thermonuclear reactions occurring inside it at incredible rates. It is estimated that the Sun converts more than 4 million tons of hydrogen to helium each second. This generates an enormous amount of heat, estimated to be about 25 million degrees Fahrenheit at the Sun's center, which reaches out to warm the Earth's surface and make life possible. The Sun's surface is cooler than its center, but still a scorching $10,000°$ Fahrenheit.

Figure 7.2: *A total solar eclipse has frequently been described as one of the most awesome astronomical events. This NASA photograph of the March 7, 1970 solar eclipse was taken at Miahuatlan, Mexico. The Moon totally blocked out the Sun for three and one-half minutes.*

Our Sun is also very bright. In terms of stellar magnitudes, the Sun is magnitude −26.8. By contrast, a Full Moon is about magnitude −12.7 while the faintest star visible to the naked eye is about magnitude 6. Because of its brightness, it is very hazardous to look directly at the Sun with the naked eye and even more dangerous to view the Sun through a telescope. The only safe way to view the Sun, even during an eclipse, is to project its image onto a screen of some sort or use specially designed filters. Staring at the Sun is a sure way to go blind.

When an image of the Sun is projected onto a screen, its surface features become more apparent. Sunspots appear on the surface as dark blotches. These dark spots only appear dark because they are cooler than their surrounding environment. Sunspots are estimated to be between 6,000° and 7,000° Fahrenheit. They are known to increase and decrease in cycles averaging about 11 years but the reason for sunspots is not completely understood.

In addition to sunspot activity, solar flares periodically erupt from the Sun's surface. These flares shoot out from the Sun with great speed and can disrupt radio communications upon the Earth. In fact, these solar flares send out showers of proton particles which reach

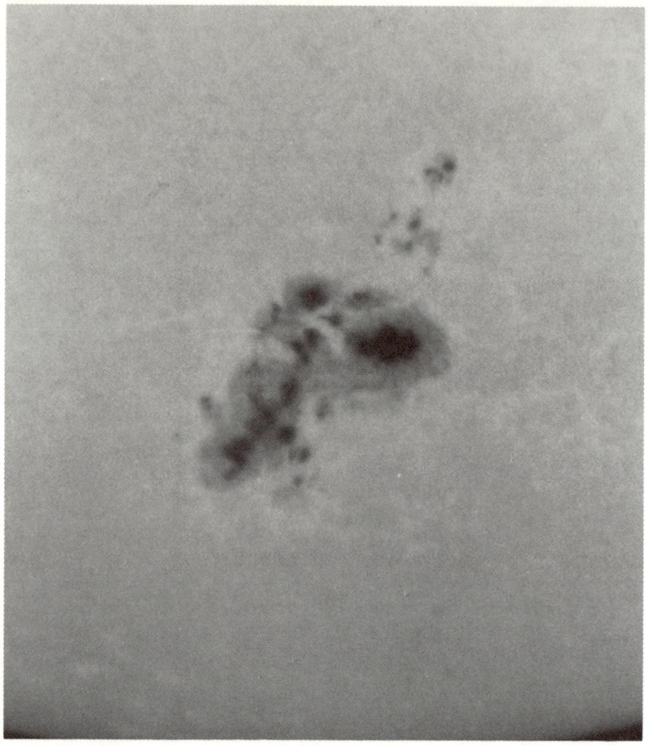

Figure 7.3: *This large group of Sunspots was taken on June 28, 1988 by Randy Tatum with a 7-inch f/15 refractor, 2× for f/30 and a Daystar T-Scanner Tuned 1–2Å to the red.*

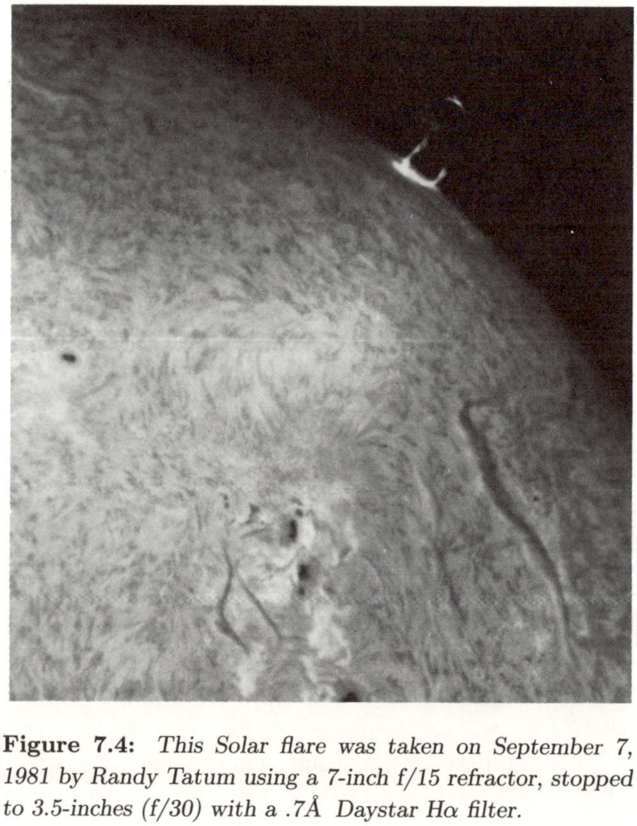

Figure 7.4: *This Solar flare was taken on September 7, 1981 by Randy Tatum using a 7-inch f/15 refractor, stopped to 3.5-inches (f/30) with a .7Å Daystar Hα filter.*

the Earth in a matter of hours and can cause disruptions in the Earth's magnetic field. This disturbance of the Earth's magnetic field in turn disrupts electromagnetic communications on the Earth's surface.

While landing on the surface of the Sun is totally out of the question, spacecraft can be sent to within a few million miles of the Sun. Such probes would greatly increase our understanding of the Sun and hence our understanding of stars.

7.1 Locating the Sun

The problem of calculating the position of the Sun, Moon, or one of the planets may seem overwhelming. Conceptually, however, the process is quite simple:

- Take a "snapshot" of the ecliptic coordinates of the object at some convenient instant in time.
- Calculate how many days (D), including fractional parts of a day, that have elapsed since the snapshot was taken.
- Calculate how far the object has moved in its orbit in D days.
- If necessary, apply corrections to account for irregularities in the object's orbit.
- Convert the corrected coordinates to horizon coordinates.

In the Sun's case, the process is simplified since only the ecliptic longitude needs to be calculated. The ecliptic latitude is always 0° because the Earth-Sun orbit lies completely within the ecliptic plane.

The process of locating a solar system object generally begins by assuming a perfectly circular orbit around the Earth. This circular orbit is called the mean orbit. The mean anomaly, angle M, is then calculated which represents how far the object has moved in degrees since some standard instant in time. From the mean anomaly, the angle v, called the true anomaly, is calculated which is how far the object has moved in degrees along its eliptical orbit. Thus, the mean anomaly essentially represents the object's ecliptic coordinates assuming a circular orbit while the true anomaly represents the object's ecliptic coordinates assuming an eliptical orbit. Locating the Sun will be considered now with the following chapters dealing with the Moon and planets.

The mean Sun is an imaginary Sun which moves in a circular orbit around the Earth. Figure 7.5 shows the mean orbit of the mean Sun superimposed on the true eliptical orbit of the Sun with the Earth pictured as the center of the mean Sun's orbit. Perigee, the moment at which the Sun is closest to the Earth, is also shown with angle M_p being measured with respect to the moment of perigee. That is, M_p is a measure of how many degrees the mean Sun has moved in a circular orbit since the moment of perigee. Since the Earth completes an orbit around the Sun in 365.2422 days, the mean anomaly is given by

$$M_p = \frac{360 * D}{365.2422} \tag{7.1.1}$$

where D is the number of days (including fractional portions of a day) which have elapsed since the moment of perigee. The subscript p is used to emphasize that the moment of perigee is used as the reference point.

There are two problems with the mean anomaly defined in this way. First, it is not the desired ecliptic longitude because the ecliptic longitude is measured with respect to the First Point of Aries ♈ while this definition of M_p is with respect to the moment of perigee. Secondly, what is really desired is the true anomaly v instead of M_p. The first difficulty will be resolved by

ϵ, eccentricity	0.016718
r_0, length of the semi-major axis	1.495985E08 km
θ_0, angular diameter when length of semi-major axis distance away	0.533128°
ε_g, ecliptic long at epoch 1980.0	278.833540°
ϖ_g, ecliptic long at perigee	282.596403°

Table 7.1 Sun's Orbital Elements. *These orbital elements represent a snapshot of the Sun's location at epoch 1980.0*

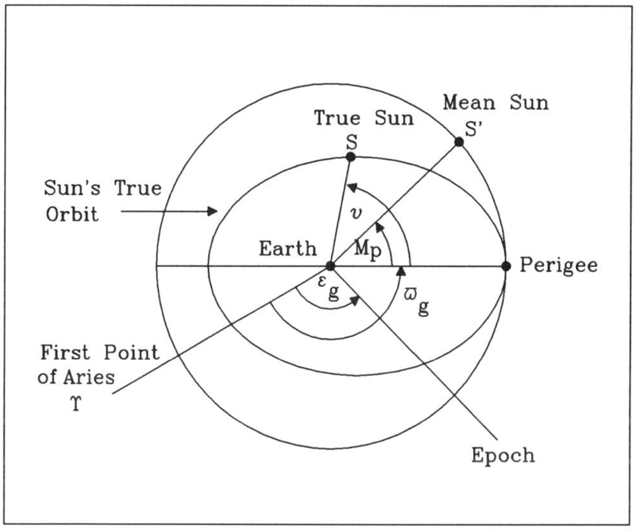

Figure 7.5: Orbit of mean Sun. *Calculating the Sun's location is simplified by assuming a circular orbit.*

choosing a new point of reference for M which will then be used to calculate v to resolve the second difficulty.

Selection of perigee as a reference point for defining M is arbitrary. Any other convenient point, such as a standard epoch, can be used as well. Figure 7.5 shows the position of the standard epoch and defines a new angle ε_g which is the angular distance from the First Point of Aries Υ to the standard epoch. Defined in this way, ε_g is the ecliptic longitude of the Sun at the standard epoch. Table 7.1 gives the value for ε_g as 278.833540° for the standard epoch 1980.0. The standard epoch 1980.0 will be used in this book as the "snapshot" moment for calculating the position of the Sun, but it should be obvious how any other standard epoch could be used.

Figure 7.5 defines another angle ϖ_g which is the ecliptic longitude of the Sun at the moment of perigee. Given ε_g and ϖ_g , a mean anomaly can then be defined which is identical to the ecliptic longitude for a circular orbit. That is, if M_p is the mean anomaly measured from perigee, then the mean anomaly measured from the First Point of Aries is

$$M = M_p + \varepsilon_g - \varpi_g \tag{7.1.2}$$

or, combining with equation 7.1.1,

$$M = \frac{360 * D}{365.2422} + \varepsilon_g - \varpi_g. \tag{7.1.3}$$

Given the value of M, Kepler's equation must be solved to compute v. Kepler's equation is given by

$$M = E - \epsilon * \sin E \tag{7.1.4}$$

where ϵ is the orbital eccentricity of the Sun-Earth orbit and M is calculated from equation 7.1.3. Once Kepler's equation has been solved for E, the true anomaly is given by

$$\tan\left(\frac{v}{2}\right) = \left(\sqrt{\frac{1+\epsilon}{1-\epsilon}}\right) * \tan\left(\frac{E}{2}\right). \tag{7.1.5}$$

To compute E from Kepler's equation requires an iterative algorithm that imposes more calculations than are warranted for the purposes of this book. Instead, the true anomaly can be approximated by using the equation of center E_c given by

$$E_c = \left(\frac{360 * \epsilon}{\pi}\right) * \sin M. \tag{7.1.6}$$

The true anomaly is then approximated by

$$v = M + E_c. \tag{7.1.7}$$

Once the true anomaly is found, the location of the Sun in ecliptic coordinates is a simple matter. The ecliptic latitude is always 0° while the ecliptic longitude is basically the true anomaly, v, except that the calculated v is measured with respect to the moment of perigee. Thus, the ecliptic longitude is given by

$$\lambda = v + \varpi_g \tag{7.1.8}$$

or equivalently

$$\lambda = M_p + E_c + \varepsilon_g. \tag{7.1.9}$$

It is worth digressing a moment to describe an ellipse since it plays such an important part in not only the Sun's orbit but the orbit of the planets as well. Figure 7.6 shows an ellipse. P is a point on the ellipse while A and B are called the foci. An ellipse is generated in such a way that the length of line segment AP plus the length of line segment BP is the same no matter where the point P is located. Point E is the center of the ellipse. The line segment CD contains both foci A and B and is called the major axis. Half of the major axis, line segments CE or ED, is called the semi-major axis.

The eccentricity ϵ is a measure of the flatness of the ellipse. Mathematically, the eccentricity is the ratio of the distance from the focus to the center to the length of the semi-major axis. In other words,

$$\epsilon = \frac{\text{length of } AE}{\text{length of } CE}. \tag{7.1.10}$$

If the focus and center of the ellipse coincide, the eccentricity is zero (because the length of AE is 0) and the resulting figure is a circle. In Figure 7.6, this implies that points A, E, and B are the same and the result is indeed a circle. The closer the value of ϵ is to 0, the closer an orbit becomes to being a circle. Table 7.1 gives the Earth-Sun orbital eccentricity as 0.016718 which shows how close the orbit is to being a circle.

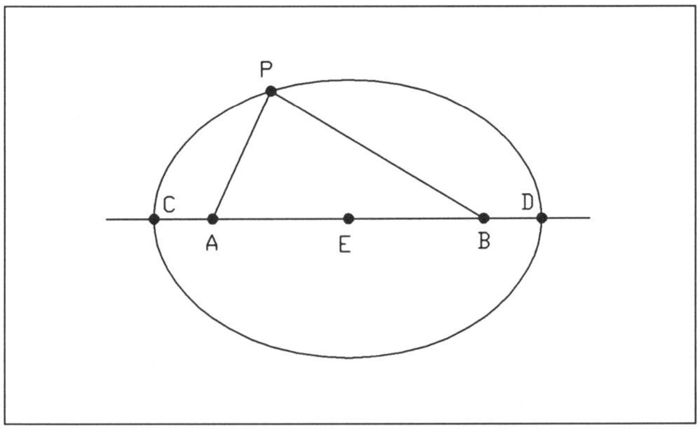

Figure 7.6: *An ellipse is defined so that the distance $AP + PB$ is constant no matter where P is located.*

Knowing the eccentricity of an ellipse and the length of the semi-major axis is required in orbital calculations.

Table 7.1 gives pertinent details relative to the Sun's orbit.

Let us find the location of the Sun on February 5, 1986 at 12^h LMT. Assume that the observer is in the Eastern Standard Time Zone at 78°W longitude, 38°N latitude. Also assume that the observer is not on Daylight Saving Time.

1. Do all the needed time conversions. That is, compute GMT, GST, and LST from the given LMT. Adjust the date if necessary.
   (Ans: $GMT = 17.0^h$, $GST = 2.036559^h$, $LST = 20.836559^h$, Date = 2/5/1986)
2. Compute the number of elapsed days since January 0.0 of the given year.
   (Ans: $Days = 36$)
3. Let T be the number of leap years since 1980.
   (Ans: $T = 2$)
4. Compute D, the total number of elapsed days since the standard epoch 1980.0 and include the fractional part of the day.
   $$D = T + 365 * (Year - 1980) + Days + \frac{GMT}{24}.$$
   It is necessary to include the fractional part of a day since the coordinates of the Sun change relatively quickly with respect to time.
   (Ans: $D = 2228.708$)
5. Compute
   $$M_p = \frac{360 * D}{365.2422}.$$
   (Ans: $M_p = 2196.720343°$)
6. Add or subtract multiples of 360° to convert M_p to the range of 0° to 360°.
   (Ans: $M_p = 36.720343°$)
7. Compute the Sun's mean anomaly with respect to ♈.
   $$M = M_p + \varepsilon_g - \varpi_g$$
   where $\varepsilon_g = 278.833540°$ and $\varpi_g = 282.596403°$.
   (Ans: $M = 32.957480°$)
8. If $M < 0°$, add 360°.
   (Ans: no adjustment)
9. Compute the equation of center.
   $$E_c = \left(\frac{360 * \epsilon}{\pi}\right) * \sin M$$
   where ϵ is the Earth's orbital eccentricity. Note that the true anomaly, v, can be computed by adding E_c to M.
   (Ans: $E_c = 1.042195°$)
10. $\lambda = M_p + E_c + \varepsilon_g$.
    (Ans: $\lambda = 316.596078°$)
11. If $\lambda > 360°$, subtract 360°.
    (Ans: no adjustment)
    At this point, the Sun's ecliptic coordinates are $\lambda°$ ecliptic longitude, 0° ecliptic latitude.
12. Convert the ecliptic coordinates to equatorial coordinates.
    (Ans: $\alpha = 21^h16^m$, $\delta = -15°52'$)
13. Finally, convert the equatorial coordinates to horizon coordinates.
    (Ans: $h = 35°47'$, $A = 172°17'$)

7.2 Sunrise and Sunset

As was the case with locating a star in horizon coordinates, there is no guarantee that the Sun will be visible in the sky. For example, the location of the Sun for the previous example at 20^h LMT is $h = -28°0'15''$, $A = 271°26'7''$ which is below the observer's horizon. However it is usually pretty easy in the Sun's case to know when it will or will not be visible. Quite simply, the Sun cannot be seen during the LMT evening hours for the observer.

Calculating the times of Sunrise and Sunset is an interesting thing to do because it illustrates very clearly that days as defined by the apparent Sun are not equal in length. An interesting experiment to perform is to plot the length of daylight hours throughout the year and observe when the days are the shortest and when they are the longest.

Determining the times for Sunrise and Sunset is not quite as easy as for determining when a star will rise or set. The reason for this is that the coordinates calculated for the Sun are good only for the time chosen. Since the Sun moves rather rapidly across the sky, its equatorial coordinates for Sunrise can differ substantially from the equatorial coordinates for Sunset. To take this into account, the Sun's position will be calculated at midnight for the date in question and then again at midnight for the next day. Given an ecliptic longitude for the Sun, the ecliptic longitude 24^h later is calculated by

$$\lambda_2 = \lambda_1 + 0.985647°. \tag{7.2.1}$$

Using these two locations (λ_2 and λ_1), the procedure from Chapter 6 will be used to compute two sets of rising and setting times, $ST1$ and $ST2$. Then an interpolation will be done using these two sets of rising and setting times to arrive at an interpolated Sunrise and Sunset time. The interpolation formula required is

$$T = \frac{24.07 * ST1}{24.07 + ST1 - ST2}. \tag{7.2.2}$$

Calculate the time of Sunrise and Sunset for February 5, 1986 for an observer at 78°W longitude, 38°N latitude. The observer is in the Eastern Standard Time Zone and is not on Daylight Saving Time.

1. Calculate the Sun's ecliptic location at midnight ($GMT = 0^{\mathrm{h}}$) for the date in question. (Ans: $\lambda = 315.878329°$, $\beta = 0°$)
2. Convert the ecliptic coordinates to equatorial coordinates. (Ans: $\alpha = 21.222580^{\mathrm{h}}$, $\delta = -16.078554°$)
3. Using the equatorial coordinates from the previous step, compute $ST1_r$ and $ST1_s$, the LST rising and setting times. (Ans: $ST1_r = 16.090182^{\mathrm{h}}$, $ST1_s = 2.354977^{\mathrm{h}}$)
4. Calculate the Sun's ecliptic location for 24^{h} later using equation 7.2.1. (Ans: $\lambda = 316.863976°$, $\beta = 0°$)
5. If (λ from step 4) $> 360°$, subtract 360°. (Ans: No adjustment required)
6. Convert the ecliptic coordinates from the previous step to equatorial coordinates. (Ans: $\alpha = 21.287779^{\mathrm{h}}$, $\delta = -15.783401°$)
7. Using the equatorial coordinates from the previous step, calculate $ST2_r$ and $ST2_s$, the LST rising and setting times. (Ans: $ST2_r = 16.138326^{\mathrm{h}}$, $ST2_s = 2.437232^{\mathrm{h}}$)
8. Interpolate the two sets of LST rising times using equation 7.2.2.
$$T_r = \frac{24.07 * ST1_r}{24.07 + ST1_r - ST2_r}.$$
(Ans: $T_r = 16.122429^{\mathrm{h}}$)
9. Interpolate the two sets of LST setting times using the same interpolation formula as the previous step.
$$T_s = \frac{24.07 * ST1_s}{24.07 + ST1_s - ST2_s}.$$
(Ans: $T_s = 2.363052^{\mathrm{h}}$)
10. Convert the LST times to their corresponding LMT times for the Sunrise and Sunset times. (Ans: $LMT_r = 7.298749^{\mathrm{h}}$, $LMT_s = 17.511415^{\mathrm{h}}$)
11. Convert the LMT times to HMS format. (Ans: $LMT_r = 7^{\mathrm{h}}18^{\mathrm{m}}$, $LMT_s = 17^{\mathrm{h}}31^{\mathrm{m}}$)

The Sunrise and Sunset times calculated will not match what is observed for the same reasons that the star rise and set times from Chapter 6 will not match observation. But atmospheric conditions cause another effect that is observable with the Sun. This effect is what we call dawn in the morning and twilight in the evening. When the Sun is less than 18° below the horizon at Sunrise, light is reflected by the atmosphere so that we see Sunlight before the Sun actually rises. The same situation occurs at Sundown when the Sun is no more than 18° below the horizon.

7.3 Equinoxes and Solstices

It has already been noted that the ecliptic plane is inclined with respect to the celestial equator. The angle of inclination is called the obliquity of the ecliptic and is about 23°26′. Twice a year the Sun crosses the celestial equator. These two moments are called the equinoxes and occur at about March 21 and September 22. The March equinox is known as the vernal equinox (or spring equinox) while the September equinox is known as the autumnal equinox (or fall equinox).

During the discussion of the ecliptic coordinate system in Chapter 5, it was noted that extending the line which is the intersection of the ecliptic plane and the celestial equator until it intersects the celestial sphere forms two points on the celestial sphere. One point is the autumnal equinox while the other is the vernal equinox. Recall that the vernal equinox (First Point of Aries, ♈) plays an important role in defining the ecliptic coordinate system and the equatorial coordinate system.

There are a couple of interesting things to note about the equinoxes. At the vernal equinox, the Sun's right ascension is 0^{h}. At the autumnal equinox, the Sun's right ascension is 12^{h}. Furthermore, at the times of the equinoxes, the length of day and night are very close to the same. This can be shown by calculating the Sunrise and Sunset times around March 21 and September 22.

The solstices occur when the Sun has no apparent Northward or Southward motion. That is, the solstices occur when the Sun is at its highest point above the plane of the celestial equator and when it is at its lowest point below the celestial equator. These two times occur at about June 22 (summer solstice) and about December 22 (winter solstice). The summer solstice is the longest day of the year while the winter solstice is the shortest day of the year. (Day here is used in the sense of the number of daylight hours.)

The solstices and equinoxes play a practical role in our calendar system. Passage through a solstice or equinox defines the beginning of another season. The March equinox marks the beginning of spring, the June solstice marks the beginning of summer, the September equinox marks the beginning of fall, and the December solstice marks the beginning of winter. Figure 7.7 illustrates the relationship of the solstices, equinoxes, celes-

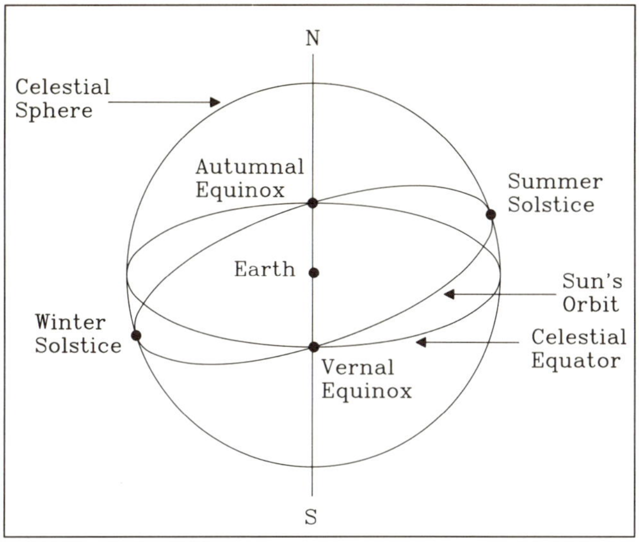

Figure 7.7: Defining the equinoxes and solstices. *The equinoxes and solstices mark the beginning and end of the seasons.*

tial equator, and the Sun's orbit. At the equinoxes and solstices, the Sun's ecliptic longitude is a multiple of 90° (0°, 90°, 180°, or 270°).

Calculating the solstices and equinoxes for a given year is a simple matter. To do so, four equations are involved for computing the Julian date. (These equations were derived by Jean Meeus and are from his book *Astronomical Formulae for Calculators*, 3rd edition, page 90.) They are as follows:

March Equinox:

$$JD = 1{,}721{,}139.2855 + 365.2421376Y + 0.0679190T^2 - 0.0027879T^3 \quad (7.3.1)$$

June Solstice:

$$JD = 1{,}721{,}233.2486 + 365.2417284Y - 0.0530180T^2 + 0.0093320T^3 \quad (7.3.2)$$

September Equinox:

$$JD = 1{,}721{,}325.6978 + 365.2425055Y - 0.1266890T^2 + 0.0019401T^3 \quad (7.3.3)$$

December Solstice:

$$JD = 1{,}721{,}414.3920 + 365.2428898Y - 0.0109650T^2 - 0.0084885T^3 \quad (7.3.4)$$

where Y is the year in question and T is the year divided by 1000. The result of each of these equations is a Julian date.

Let us calculate the solstices and equinoxes for 1985. The equations above are accurate to about 15 minutes although the calculations below will be given to the nearest second.

1. Calculate
$$T = \frac{Year}{100}.$$
(Ans: $T = 1.985$)
2. Using the equations above, compute the Julian Dates.
(Ans: March $JD = 2{,}446{,}145.1744$,
June $JD = 2{,}446{,}237.9436$,
September $JD = 2{,}446{,}331.5872$,
December $JD = 2{,}446{,}421.4187$)
3. Convert the Julian dates, JD, to calendar dates and times.
(Ans: March JD gives 3/20.67/1985,
June JD gives 6/21.44/1985,
September JD gives 9/23.09/1985,
December JD gives 12/21.92/1985)
4. Convert the day part of the dates to day and GMT.
(Ans: March: day 20, 16.186707^h,
June: day 21, 10.645432^h,
September: day 23, 2.092987^h,
December: day 21. 22.047775^h)
5. Convert the GMT results to HMS format and adjust the date.
(Ans: March equinox: 3/20/1985 at 16^h11^m,
June solstice: 6/21/1985 at 10^h39^m,
September equinox: 9/23/1985 at 2^h6^m,
December solstice: 12/21/1985 at 22^h3^m)

7.4 Solar Distance and Angular Size

The Sun is about 93 million miles away. But its distance from the Earth varies throughout the year because the orbit of the Earth is elliptical in shape. If the orbit was a perfect circle, the Earth-Sun distance would be constant.

At first glance, it might seem that the fact that the Earth-Sun distance varies throughout the year accounts for the seasons. It is true that distance from the Sun is an important factor in determining how much heat reaches a planet's surface. Mercury is very hot since its orbit is so close to the Sun. Pluto's surface is very cold since its orbit is so far away from the Sun. However, the Earth's seasons are more attributable to the inclination of the axis of rotation than to the varying Earth-Sun distance.

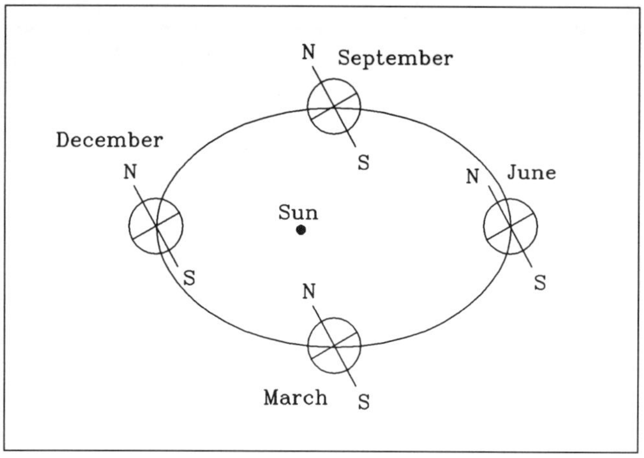

Figure 7.8: Changing seasons. *Our seasons are more attributable to the tilt of the Earth's axis than the varying Earth-Sun distance.*

Figure 7.8 illustrates the relationship of the axis of rotation to the Earth's orbit. In the center of this greatly exaggerated picture is the Sun. The Earth is shown orbiting around the Sun in an ellipse. The line extending through the Earth is the Earth's axis of rotation. Instead of the axis of rotation being perpendicular to the plane of the Earth's orbit, it is actually inclined away from the perpendicular by an angle of about 23°26′. This is the angle called the obliquity of the ecliptic.

Since the Earth's axis of rotation is not perpendicular to the Earth's orbital plane, the Sun shines more directly on the Northern Hemisphere during one portion of the year (summer) than six months earlier or later (winter) and the days are longer in the summer than in the winter. The result is that more Sunlight reaches the Northern Hemisphere in the summer and these direct rays are more concentrated, which results in warmer weather than in winter when the Sun's energy strikes the ground at a more slanted, or less concentrated, angle. Figure 7.8 also shows why a particular part of the year experiences opposite seasons in opposite hemispheres. The slant of the Earth's axis is the true cause of the seasons; the Earth's varying distance from the Sun has far less effect.

Since the distance to the Sun varies throughout the year, the Sun's apparent size also varies. Once the true anomaly is known, the distance to the Sun and the Sun's angular size can be calculated quite easily. The equations required are

$$Dist = \frac{r_0 * (1 - \epsilon^2)}{1 + \epsilon * \cos \upsilon}$$

and

$$\theta = \frac{\theta_0 * (1 + \epsilon * \cos \upsilon)}{1 - \epsilon^2}$$

where r_0 is the length of the semi-major axis of the Earth-Sun orbit and θ_0 is the angular diameter of the Sun when it is r_0 distance away.

To illustrate, find the distance to the Sun and the Sun's angular size on February 15, 1985.

1. Using the method of the section on computing the Sun's location, compute the Sun's true anomaly for the given date (at 0^h GMT). Recall that the true anomaly can be found by adding E_c to the mean anomaly, M.
   (Ans: $\upsilon = 43.645258°$)
2. Compute
   $$F = \frac{1 + \epsilon * \cos \upsilon}{1 - \epsilon^2}$$
   where $\epsilon = 0.016718$ is the Earth's orbital eccentricity.
   (Ans: $F = 1.012381$)
3. Compute the distance
   $$Dist = \frac{r_0}{F}$$
   $r_0 = 1.495985\text{E}08$ km is the length of the Earth's semi–major orbital axis.
   (Ans: $Dist = 1.478\text{E}08$ km)
4. $\theta = F * \theta_0$ where θ_0 is the Sun's angular diameter when it is r_0 km away. $\theta_0 = 0.533128°$.
   (Ans: $\theta = 0.5397°$)
5. Convert distance to miles and angular diameter to DMS if desired.
   (Ans: $Dist = 9.182\text{E}07$ miles, $\theta = 0°32'$)

7.5 Equation of Time

During the discussion of time in Chapter 4, it was pointed out that the Sun can be used to define a day. This day is called the solar day. However, the motion of the Sun is irregular. For this reason a mean Sun is defined from which another definition of day can be obtained. This day is called the mean solar day. Our wristwatches tell time according to a mean solar day.

The amount by which time for a mean solar day and time for a solar day differ varies throughout the year. The difference between the two is called the equation of time. If the solar day time is known (perhaps by a sundial) and the equation of time is known, the mean solar time can be easily calculated. This is done by adding the solar time to the equation of time. Mathematically, the relationship between solar time, mean solar time, and the equation of time, ΔT, is

$$\Delta T = Mean\ solar\ time - Solar\ time. \quad (7.5.1)$$

Finding the equation of time is straightforward. By definition the mean Sun transits the observer's horizon at exactly 12^h. All that is needed now is to find the equivalent solar time at the same instant. This can be done by recalling that the right ascension and sidereal time are the same at transit. Therefore, finding the Sun's right ascension at transit will give the GST from which the GMT can be calculated. The equation of time will then be the difference between 12^h and the calculated GMT.

An alternative way to calculate the equation of time is to compute it directly from the Sun's ecliptic longitude and mean anomaly. The required equation, given by W. M. Smart in *Textbook on Spherical Astronomy*, is

$$\Delta T = y * \sin(2 * \lambda) - 2 * \epsilon * \sin M + 4 * \epsilon y * \sin M * \cos(2 * \lambda) - \frac{1}{2} * y^2 * \sin(4 * \lambda) - \frac{5}{4} * \epsilon^2 * \sin(2 * M) \quad (7.5.2)$$

where

$$y = \tan^2 \frac{\varepsilon}{2}.$$

Table 7.1 gives the Earth's orbital eccentricity, ϵ , while an acceptable average value for the obliquity of the ecliptic, ε , is 23.441884°. The resulting equation of time given by this formula is in radians which must be converted to degrees and then to hours.

To illustrate, consider calculating the equation of time for May 5, 1986.

1. Compute the Sun's ecliptic longitude and mean anomaly for the date given.
(Ans: $\lambda = 44.722318°$, $M = 120.474829°$)
2. Compute y.
(Ans: $y = 0.043044$)
3. Using equation 7.5.2, compute ΔT in radians.
(Ans: $\Delta T = 0.014537$radians)
4. Convert radians to degrees.
(Ans: $\Delta T = 0.832904°$)
5. Convert degrees to hours by dividing by 15.
(Ans: $\Delta T = 0.055527^h$)
6. Convert the equation of time to HMS format if desired.
(Ans: $\Delta T = 0^h3^m20^s$)

In this example, 3^m20^s must be added to the sundial time to convert it to the mean solar time.

7.6 About the Program

The program for this chapter is SUN which contains the algorithms described in this chapter. If you examine the subroutines, you will note that many of them come from preceding chapters. The subroutines included in this program calculate the location of the Sun, the times for Sunrise and Sunset, the times for the equinoxes and solstices, and the equation of time.

7.7 Sample Problems

1. An observer is located at 95°W longitude, 30°N latitude within the Central Standard Time Zone. The date is August 9, 1980 and the observer is on Daylight Saving Time. If the LMT is 12^h, what are the Sun's ecliptic coordinates? What are the Sun's equatorial coordinates? What are the Sun's horizon coordinates?
(Ans: $\beta = 0°$, $\lambda = 137°14'$,
$\alpha = 9^h19^m$, $\delta = 15°40'$,
$h = 65°45'$, $A = 121°28'$)
2. What are the times for Sunrise and Sunset for the previous observer?
(Ans: $LMT_r = 6^h46^m$, $LMT_s = 20^h5^m$)
3. Another observer is located at 30°W longitude, 20°S latitude. He is in the Eastern Standard Time Zone and is on Daylight Saving Time. If the date is May 6, 1984 and LMT is $14^h30^m0^s$, what are the Sun's ecliptic coordinates? What are the Sun's equatorial coordinates? What are the Sun's horizon coordinates?
(Ans: $\beta = 0°$, $\lambda = 47°23'$,
$\alpha = 3^h0^m$, $\delta = 17°2'$,
$h = 14°13'$, $A = 294°26'$)
4. What are the times for Sunrise and Sunset for the previous observer?
(Ans: $LMT_r = 4^h28^m$, $LMT_s = 15^h33^m$)
5. For the year 1984, what are the GMT times for the solstices and equinoxes?
(Ans: March Equinox: March 20 at 10^h22^m,
June Solstice: June 21 at 4^h51^m,
September Equinox: September 22 at 20^h17^m,
December Solstice: December 21 at 16^h31^m)
6. On August 9, 1980, how far away was the Sun? What was its angular diameter?
(Ans: *Dist* = 1.517E08 km or 9.424E07 miles, $\theta = 0°32'$)
7. On May 6, 1984, how far away was the Sun? What was its angular diameter?
(Ans: *Dist* = 1.510E08 km or 9.381E07 miles, $\theta = 0°32'$)
8. On August 9, 1980, what was the equation of time?
(Ans: $\Delta T = -0^h5^m28^s$)
9. On May 6, 1984, what was the equation of time?
(Ans: $\Delta T = 0^h3^m28^s$)

10. On January 1, 2000, what will be the equation of time?
(Ans: $\Delta T = -0^{h}3^{m}49^{s}$)

Chapter 8

The Moon

This chapter is devoted to calculations involving Earth's only natural satellite—the Moon. Our Moon is by far the most visible of all the celestial objects in the night sky. And unlike most other celestial objects, the Moon is quite often visible during daytime hours. Its domination of the night sky and its quiet beauty has caused the Moon to be a source of wonder throughout the ages. Even with just the naked eye the Moon is an interesting object to behold. But the enjoyment of its splendor is greatly enhanced by a low powered telescope or even a set of binoculars.

This chapter will demonstrate how to calculate the position of the Moon in terms of ecliptic, equatorial, and horizon coordinates. Such calculations may seem to be more trouble than they are worth since the location of the Moon can be readily determined by just looking up into the night sky. However, calculating the Moon's position is required for predicting the occurrence of lunar and solar eclipses. Furthermore, calculating the times of Moonrise and Moonset require knowing the Moon's location. Moonrise and Moonset times will be calculated while the calculations for eclipses will only be described in general terms.

Besides determining the location of the Moon, this chapter will also provide algorithms for calculating the distance to the Moon and the angular size of the Moon. The angular size of the Moon is close to that of the Sun. This is because the Moon is so much closer to the Earth than the Sun is and not because the Moon is nearly the same size as the Sun. In fact, the Moon is only about a fourth the size of the Earth and is thus greatly dwarfed in comparison to the Sun.

As explained in Chapter 4, the definition of a month is closely tied to the Moon. The division of the month into roughly four weeks is historically related to the four phases of the Moon. Calculations will be presented to show how the phase of the Moon can be determined. But before delving into any of the calculations, a brief description of the Moon is in order.

Fascination with the Moon has inspired science fiction writers to create fanciful tales of lunar creatures or daring space flights. Perhaps most notable was the French author Jules Verne (1828–1905) who wrote of an imaginary lunar trip with surprising accuracy. His approach to sending a man to the Moon is not all that far from what our modern day rockets do. Jules Verne envisioned a spacecraft being shot out of a gigantic cannon which may be a fair description of how the astronauts feel when they are blasted into space. The description of the state of weightlessness on such an imaginary flight gives an indication of how prophetic Jules Verne has turned out to be.

Figure 8.1: *This NASA photograph of the Moon was taken from the Apollo 11 spacecraft during its journey homeward. When this picture was taken the spacecraft was already 10,000 nautical miles away. Aboard Apollo 11 were astronauts Neil A. Armstrong, Michael Collins, and Edwin E. Aldrin, Jr.*

There has long been a belief that the Moon, and especially a full Moon, affects the behavior of humans. The classic legend of the werewolf is one example of a belief

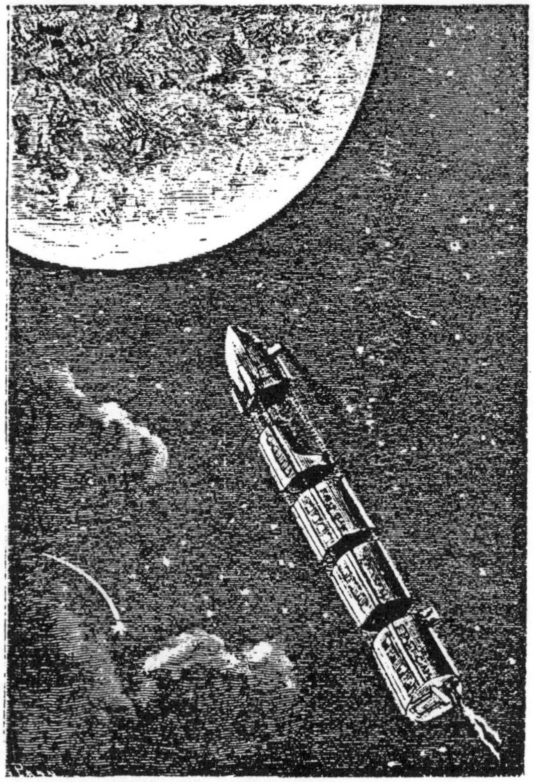

Figure 8.2: *The above drawing from* From the Earth to the Moon *was drawn almost 100 years before the Space Age. Jules Verne's fantasy has become today's reality.*

dating back to the medieval period. Our word *lunatic* dates back at least that far and stems from an early belief that the Moon could cause insanity. But interestingly enough, some modern day "researchers" have suggested a correlation between the phase of the Moon and the number of violent crimes which are committed during a month. (Such research is difficult to take seriously!) Who knows? Maybe the old legends are not so wild after all ....

The Moon has a more direct effect on the Earth than simply fueling our imaginations and providing romantic Moonlight. The tides are caused by the gravitational pull of the Sun and Moon upon the oceans and are readily visible effects to anyone that lives near an ocean. Moreover, the gravitational pull of the Moon affects the Earth's orbit just as the gravitational pull of the Earth affects the Moon's orbit.

Features on the surface of the Moon serve to enhance its romantic attraction. The silver color contrasted with dark patches here and there have caused some to look up at the full Moon and visualize the face of a man. Others deny that there is a "man in the Moon" and have visualized the profile of a woman instead. Still others in times past imagined great oceans on the Moon as evidenced by the dark areas on its surface. In fact, the

Figure 8.3a and b: *These two photographs were taken at "The Rocks" on Hopewell Cape in the Bay of Fundy. While the geography of the Bay of Fundy contributes to the extreme movement of the tides, the Moon's gravitational pull is the force that sets this awesome power in motion. Photographs courtesy Tourism New Brunswick, Canada.*

dark areas of the Moon are called *maria* which is the Latin word for seas.

Studies of the Moon made in this century have dispelled many of the romantic notions once held about the Moon. Lunar features are easily recognizable with even a moderately sized telescope and because of this, much of the mystery is gone. This is most notably true with the giant craters which can be seen all across the surface in great profusion, perhaps denoting a violent past when the Moon was bombarded by meteorites. Visible to the naked eye, the crater Tycho gives mute evidence of the tremendous size of the meteorites that must have struck the Moon in ages past. Further to the North, the giant crater Copernicus is nearly 40 miles wide and 12,000 feet deep. Tycho is also of interest because of the clearly discernible rays which emanate from around the rim of the crater.

Whether or not water was ever on the Moon may be debatable, but it is certain that there are no massive lunar seas. Satellite photographs and actual manned expeditions to the Moon confirm that the great lunar maria are not seas at all. They are in fact relatively flat plains.

Another striking feature of the lunar surface is that the Moon contains a number of very large mountain ranges. The most mountainous region visible from Earth is in the Moon's Southern hemisphere. Located near the South Pole, the Leibnitz Mountains have peaks up to 19,600 feet high. While not as tall as Mount Everest, the Leibnitz Mountains are comparable in height to Mount Kilimanjaro. Other mountainous regions include the lunar Caucasian Mountains which tower to 19,000 feet and the lunar Alps which reach a height of 12,000 feet.

A curious thing about the Moon is that its rotation is such that only one side of the Moon is ever visible from the Earth. The unseen side of the Moon, erroneously called the dark side, was first photographed by the Soviets and later seen by members of the American Apollo Moon missions. Satellite photographs of the far side of the Moon reveal a difference in the two sides. Almost 90% of the far side of the Moon is covered in craterous regions while only about 60% of the side visible from Earth is covered in craters.

As stated earlier, the Moon is the only natural satellite that orbits the Earth. However, our Moon is not the only moon in the solar system. Most of the planets (all except Mercury and Venus) are known to have moons. Perhaps observing the motion of the moons of Jupiter convinced Galileo that the Sun is the center of our solar system. But as Isaac Asimov once observed, our very own Moon may have been a factor for delaying widespread acceptance of a heliocentric solar system. Since the Moon obviously orbits around the Earth, it could be argued, then so must all the other objects within the solar system and the universe.

The Moon is roughly 240,000 miles away from the Earth. It is about one fourth the size of the Earth with a diameter of some 2,160 miles. Its mass is only 0.01228 times that of the Earth. Because it has such a relatively low mass, its gravitational field is too weak to hold an atmosphere. Hence, the airless Moon is a place where there is no sound because there is no atmosphere to carry the sound waves. Shadows on the Moon are very dark and sharply defined because there is no atmosphere to scatter the light rays. There is no weather to alter the landscape so that the Moon's surface remains unchanged except for the effects of occasional Moonquakes, meteorites, or visits by man.

The lack of an atmosphere also results in very hot lunar days and very cold lunar nights. Temperatures at night may drop as low as $-300°$ Fahrenheit while daytime temperatures may soar to 250° Fahrenheit. A lunar day with its searing heat lasts for 14 Earth days while the bitterly cold nights are also about 14 Earth days in length. Even with such extremes in temperature, a lunar base is possible, especially if part of the base is under the Moon's surface. Underground homes on Earth are a rarity but could well be the norm on the lunar surface when colonization of the Moon becomes a reality.

8.1 Lunar Exploration

It would be unfair to claim that exploration of the Moon did not begin until the advent of the space age. But it is fair to say that the most exciting exploration of the Moon began when the Americans and Soviets sent spacecraft to the Moon. Of course the most climatic moment came during the Apollo 11 mission when Neil Armstrong stepped out onto the alien lunar surface.

The first missions to the Moon were all unmanned. In fact, most of the early satellite probes sent to the Moon were intentionally crashed onto the lunar surface. The first of these was the Luna 2 satellite launched by the Soviets. Luna 2 crashed onto the Moon on September 13, 1959. Shortly thereafter, Luna 3 was launched and successfully sent back photographs of the far side of the Moon on October 4, 1959 before it too crashed onto the Moon.

Following the success of the Soviet Luna program, the Americans launched a series of Ranger satellites to explore the lunar surface. The Ranger series, launched in the early 1960s, was designed to send back lunar photographs before crashing into the Moon. Surveyor satellites were launched between 1966 and 1968 to actually land on the Moon and send back important scientific data in preparation for a manned mission. But once again, the Soviets were first with Luna 9 which soft

Figure 8.4: *Astronaut David R. Scott saluting the U.S. flag. Part of the Lunar Excursion Module can be seen on the right. Hadley Mountain in the background is about 5km away. Hadley Delta, behind where Astronaut Scott is standing, rises about 4,000 meters above the plain. NASA photograph.*

landed on the Moon in 1966. Pictures sent back from Luna 9 were the first to be sent back from the surface of another world.

The pinnacle of the American space program was reached during the Apollo lunar missions. The Apollo 11 crew members Neil Armstrong, Edwin "Buzz" Aldrin, and Michael Collins succeeded in accomplishing a feat that man has dreamed about since time immemorial. On July 20, 1969 Neil Armstrong stepped onto the lunar surface while the entire world watched in awe and wonder. It is doubtful that the excitement of that moment will be realized again until a manned mission succeeds in landing on a planet within the solar system.

8.2 Locating the Moon

Calculating the position of the Moon is the most ambitious calculation presented so far in this book. Determining the position of the planets will be the only computation which is longer. This complexity is due to the fact that both the effect of the Sun's gravitational pull and the effect of the Earth's gravitational pull must be

ϵ, eccentricity	0.054900
r_0, length of the semi-major axis	384,401 km
θ_0, angular diameter when length of semi-major axis distance away	0.5181°
ι, inclination of lunar orbit to the ecliptic	5.145396°
λ_0, ecliptic long at epoch 1980.0	64.975464°
ϖ_0, ecliptic long at perigee for epoch 1980.0	349.383063°
Ω_0, ecliptic long for ascending node at epoch 1980.0	151.950429°

Table 8.1 Moon's Orbital Elements. *These orbital elements are for epoch 1980.0.*

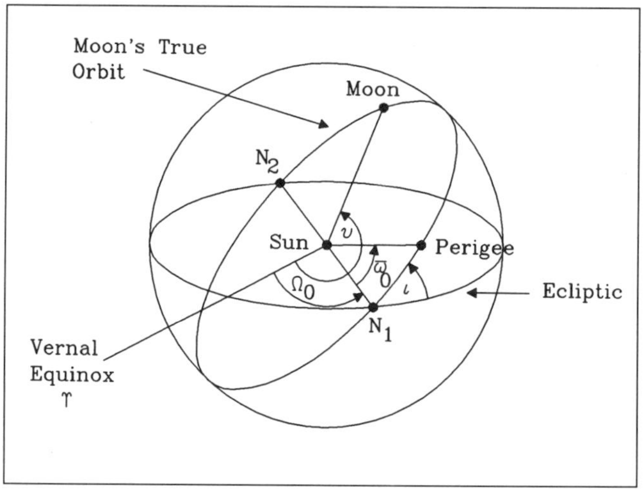

Figure 8.5: Locating the Moon. *Calculating the Moon's position is a long, tedious process!*

considered in finding the Moon's position.

Figure 8.5 shows the angles required for locating the Moon and is reminiscent of Chapter 5's figure in the section on orbital elements. The nomenclature from Table 8.1 has been used in the figure to make the items from the table more meaningful. Notice that Figure 8.5 includes the position of the Sun since it has such an impact on calculating the Moon's position.

To determine the Moon's position, we will proceed conceptually in much the same manner as for determining the Sun's location. The whole idea is to be able to calculate the Moon's true anomaly from which the Moon's position can be determined. However, locating the Moon is complicated in two ways. First, locating the

Moon is a three dimensional problem since the Moon's orbit does not lie within the ecliptic plane. Since the Sun's orbit lies completely within the ecliptic plane, only the ecliptic longitude must be calculated for the Sun. However, both the ecliptic longitude and the ecliptic latitude must be calculated for the Moon. Secondly, four major adjustments must be made in calculating the Moon's position:

- Annual equation A_e
- Variation correction V
- Evection correction E_v
- Corrected anomaly C_a.

The annual equation correction accounts for the fact that the Earth-Sun distance varies throughout the year and hence affects the gravitational pull of the Sun on the Moon. The variation correction makes an adjustment for the gravitational pull of the Sun which varies with the Moon's location. When the Moon is between the Earth and Sun, the gravitational attraction is stronger. When the Earth is between the Moon and the Sun, the gravitational attraction is weaker. The evection correction is made to account for slight changes in the value of the Moon's orbital eccentricity. Finally, corrections are applied to the mean anomaly to arrive at a corrected mean anomaly for use in later calculations.

There are other corrections which could be made to the Moon's position and indeed must be made for very accurate calculations. In fact, there are hundreds of corrections to be made to account for gravitational effects of the planets, parallax, and variation in the longitude of the ascending node and position of perigee. Fortunately, sufficient accuracy can be maintained without considering all of the theoretically important corrections. The accuracy given here is better than a quarter of a degree.

Summarizing the approach, the position of the Moon is calculated as follows:

- Using the method of Chapter 7, calculate the position of the Sun, $\lambda_\odot$.
- Compute the Moon's mean ecliptic longitude for epoch 1980.0, λ.
- Compute the Moon's mean ecliptic longitude for the ascending node, Ω.
- Compute the Moon's mean anomaly, M_m.
- Compute corrections A_e, E_v, and V.
- Apply the corrections to Ω and M_m from which the ecliptic coordinates can be obtained.

Several equations are required for this lengthy process. The major equations are summarized here and expanded upon in the example below. The subscript m will refer to the Moon while the subscript $\odot$ refers to the Sun.

The Moon's mean ecliptic longitude before any corrections are made is

$$\lambda = 13.1763966 * D + \lambda_0 \qquad (8.2.1)$$

where D is the number of days (including fractional days) since the epoch 1980.0 and λ_0 is the Moon's ecliptic longitude at epoch 1980.0. The mean ecliptic longitude of the ascending node is

$$\Omega = \Omega_0 - 0.0529539 * D \qquad (8.2.2)$$

where Ω_0 is the Moon's ecliptic longitude of the ascending node at epoch 1980.0. Recall that the ascending node is the point in the Moon's orbit at which it transitions from being below the plane of the ecliptic to being above the plane of the ecliptic. The uncorrected mean anomaly is given by

$$M_m = \lambda - 0.114041 * D - \varpi_0 \qquad (8.2.3)$$

where ϖ_0 is the Moon's ecliptic longitude at the moment of perigee for epoch 1980.0. Four corrections (evection correction E_v, annual equation correction A_e, variation correction V, and corrected mean anomaly C_a) are required as follows:

$$\begin{aligned} E_v &= 1.2739 * \sin[2 * (\lambda - \lambda_\odot) - M_m] && (8.2.4)\\ A_e &= 0.1858 * \sin M_\odot && (8.2.5)\\ C_a &= M_m + E_v - A_e - 0.37 * \sin M_\odot && (8.2.6)\\ V &= 0.6583 * \sin[2 * (\lambda' - \lambda_\odot)] && (8.2.7) \end{aligned}$$

where λ' is a corrected ecliptic longitude that includes the A_e and E_v corrections. That is,

$$\begin{aligned} \lambda' &= \lambda + E_v + 6.2886 * \sin C_a - && (8.2.8)\\ &\quad A_e + 0.214 * \sin(2 * C_a). \end{aligned}$$

Applying the variation correction, V, to λ' results in the true ecliptic longitude λ_t given by

$$\lambda_t = \lambda' + V. \qquad (8.2.9)$$

The corrected ecliptic longitude of the ascending node is given by

$$\Omega' = \Omega - 0.16 * \sin M_\odot. \qquad (8.2.10)$$

From the values of Ω', λ_t, and the Moon's orbital inclination ι , the ecliptic coordinates of the Moon are given by

$$\lambda_m = \Omega' + \tan^{-1}\left[\frac{\sin(\lambda_t - \Omega') * \cos\iota}{\cos(\lambda_t - \Omega')}\right] \qquad (8.2.11)$$

for the ecliptic longitude and

$$\sin\beta_m = \sin(\lambda_t - \Omega') * \sin\iota. \qquad (8.2.12)$$

for the ecliptic latitude. The resulting ecliptic coordinates can then be converted to equatorial or horizon as desired.

In every previous calculation, the difference between ET and GMT has been ignored. Ephemeris Time (ET) was discussed in Chapter 4. For the purposes of calculating the Moon's position, the time difference between ET and GMT can be significant. The difference in the two times varies but will be approximated here as being 50 seconds. Thus,

$$ET = GMT + 50^{s}. \tag{8.2.13}$$

To illustrate this lengthy process, let us determine the Moon's position for January 1, 1985 at 22^{h} LMT. Assume the observer is at 78°W longitude, 38°N latitude within the Eastern Standard Time Zone. Also assume that the observer is not on Daylight Saving Time. (Remember that the subscript m below represents a value for the Moon while $\odot$ represents a value for the Sun.)

1. Convert LMT to GMT, GST, and LST times and adjust date if needed.
(Ans: $GMT = 3.00^{h}$, $GST = 9.780008^{h}$, $LST = 4.580008^{h}$, Date $= 1/2/1985$)
2. Compute $ET = GMT + 50^{s}$.
(Ans: $ET = 3.013889^{h}$)
3. Compute d, the number of days since January 0.0 for the given year. Subtract Julian dates to arrive at d.
(Ans: $d = 2$ days)
4. Calculate
$$Days = d + \frac{ET}{24}.$$
This gives the total number of days including the fractional part of a day.
(Ans: $Days = 2.125579$ days)
5. Compute T, the number of leap years since the start of 1980.
(Ans: $T = 2$)
6. Compute
$$D = T + 365 * (Year - 1980) + Days.$$
where D is the total number of days since 1980.0.
(Ans: $D = 1{,}829.125579$)
7. Calculate the Sun's ecliptic longitude and mean anomaly for the given date and time using the algorithm from Chapter 7.
(Ans: $\lambda_{\odot} = 281.676527°$, $M_{\odot} = 359.109884°$)
8. $\lambda = 13.1763966 * D + \lambda_0$ where λ_0 is the mean longitude at epoch 1980.0. $\lambda_0 = 64.975464°$. λ is the Moon's mean ecliptic longitude before any corrections are applied.
(Ans: $\lambda = 24166.259520°$)
9. Add or subtract multiples of 360 to put λ into the range of 0° to 360°.
(Ans: $\lambda = 46.259520°$)
10. $M_m = \lambda - 0.1114041 * D - \varpi_0$ where ϖ_0 is the Moon's mean ecliptic longitude at perigee for epoch 1980.0 and M_m is the Moon's mean anomaly before any corrections are made. $\varpi_0 = 349.383063°$.
(Ans: $M_m = -506.895632°$)
11. If necessary, adjust M_m to the range $[0°, 360°]$.
(Ans: $M_m = 213.104368°$)
12. $\Omega = \Omega_0 - 0.0529539 * D$ where Ω_0 is the Moon's mean ecliptic longitude at the ascending node for epoch 1980.0. $\Omega_0 = 151.950429°$.
(Ans: $\Omega = 55.091096°$)
13. If necessary, adjust Ω to the range $[0°, 360°]$.
(Ans: no adjustment needed)
14. $C = 2 * (\lambda - \lambda_{\odot})$.
(Ans: $C = -470.834014°$)
15. $E_v = 1.2739 * \sin(C - M_m)$.
E_v is the evection correction.
(Ans: $E_v = 0.749886°$)
16. $A_e = 0.1858 * \sin M_{\odot}$.
This is the annual equation correction.
(Ans: $A_e = -0.002886°$)
17. Compute the corrected anomaly, C_a
$$C_a = M_m + E_v - A_e - 0.37 * \sin M_{\odot}.$$
(Ans: $C_a = 213.862888°$)
18. $E_c = 6.2886 * \sin C_a$. This value is a correction to the true anomaly.
(Ans: $E_c = -3.504056°$)
19. $A = 0.214 * \sin(2 * C_a)$.
(Ans: $A = 0.198031°$)
20. $\lambda' = \lambda + E_v + E_c - A_e + A$. This equation applies the correction terms calculated so far to arrive at a corrected ecliptic longitude.
(Ans: $\lambda' = 43.706267°$)
21. $V = 0.6583 * \sin[2 * (\lambda' - \lambda_{\odot})]$. This is the variation correction.
(Ans: $V = -0.591975°$)
22. $\lambda_t = \lambda' + V$. Apply the variation correction to the ecliptic longitude to calculate the true ecliptic longitude λ_t.
(Ans: $\lambda_t = 43.114292°$)
23. $\Omega' = \Omega - 0.16 * \sin M_{\odot}$. This results in a corrected ecliptic longitude of the ascending node, Ω'.
(Ans: $\Omega' = 55.093581°$)

24. $y = \sin(\lambda_t - \Omega') * \cos \iota$ where $\iota = 5.145396°$ is the inclination of the Moon's orbit.
(Ans: $y = -0.206722$)
25. $x = \cos(\lambda_t - \Omega')$.
(Ans: $x = 0.978223$)
26. Compute
$$T = \tan^{-1}\left(\frac{y}{x}\right).$$
(Ans: $T = -11.932403°$)
27. Using the signs of y and x, compute a quadrant adjustment for T to remove the angle ambiguity. This value will be used to calculate the Moon's ecliptic longitude λ_m which must be in the range 0° to 360°.
(Ans: $Adjustment = 360°$, $T = 348.067597°$)
28. $\lambda_m = \Omega' + T$.
(Ans: $\lambda_m = 403.161178°$)
29. If $\lambda_m > 360°$, then subtract 360°.
(Ans: $\lambda_m = 43.161178°$)
30. $Lat = \sin(\lambda_t - \Omega') * \sin \iota$.
(Ans: $Lat = -0.018615$)
31. $\beta_m = \sin^{-1} Lat$.
(Ans: $\beta_m = -1.066596°$)
32. Convert the Moon's Ecliptic Longitude, λ_m, and the Moon's Ecliptic Latitude, β_m, to their corresponding equatorial coordinates.
(Ans: $\alpha_m = 2.736060^h$, $\delta_m = 14.773809°$)
33. Convert the equatorial coordinates to horizon coordinates.
(Ans: $h_m = 56°18'$, $A_m = 233°59'$)

The calculations just presented are long indeed. Care must be taken at each step to avoid making mistakes in such a lengthy process!

8.3 Moonrise and Moonset

As in the case of the Sun, the Moon's equatorial coordinates change rather rapidly with time. Hence, to compute the times for Moonrise and Moonset, the position of the Moon will be calculated for two different times and then an interpolation will be done to determine the times of Moonrise and Moonset.

In the algorithm presented here, the Moon's position will first be determined for midnight ($GMT = 0^h$) of the date in question. Using the equatorial coordinates obtained for midnight, a set of rising and setting times will be computed. Then the location of the Moon 12 hours later will be calculated to arrive at a second set of rising and setting times. If t is the number of hours between two times, the Moon's new ecliptic coordinates t hours later are:

$$\beta_2 = \beta_1 + 0.05 * \cos(\lambda_t - \Omega') * t \quad (8.3.1)$$
$$\lambda_2 = \lambda_1 + (0.55 + 0.06 * \cos C_a) * t. \quad (8.3.2)$$

The two sets of rising and setting times will then be interpolated by the following formula:

$$T = \frac{12.03 * ST1}{12.03 + ST1 - ST2}. \quad (8.3.3)$$

As an example, consider computing the rising and setting times for the observer from the previous section.

1. Calculate the Moon's ecliptic coordinates for midnight. That is, at 0^h GMT.
(Ans: $\lambda_r = 29.715063°$, $\beta_r = -2.215239°$)
2. Let α_r and δ_r be the equatorial coordinates for the ecliptic coordinates calculated in the previous step.
(Ans: $\alpha_r = 1.895262^h$, $\delta_r = 9.298787°$)
3. $D_b = 0.05 * \cos(\lambda_t - \Omega') * t$ where λ_t and Ω' were calculated in the first step. D_b is a correction to the Moon's ecliptic latitude for t hours later. (This is much shorter than recalculating the Moon's position!) Thus, $t = 12$ for this equation.
(Ans: $D_b = 0.541411°$)
4. $D_l = (0.55 + 0.06 * \cos C_a) * t$ where C_a is the Corrected Anomaly which was calculated in the first step. D_l is a correction for the Moon's ecliptic longitude for t hours later.
(Ans: $D_l = 5.918948°$)
5. Using D_b and D_l, compute a new ecliptic longitude and ecliptic latitude for the Moon's position 12 hours later. Thus,
$$\lambda_s = \lambda_r + D_l \quad \text{and} \quad \beta_s = \beta_r + D_b.$$
(Ans: $\lambda_s = 35.634011°$, $\beta_s = -1.673828°$)
6. Compute α_s and δ_s, the equatorial coordinates for the ecliptic coordinates computed in the previous step.
(Ans: $\alpha_s = 2.259987^h$, $\delta_s = 11.822104°$)
7. Using α_r and δ_r, compute $ST1_r$ and $ST1_s$ which will be the LST rising and setting times for these equatorial coordinates. This calculation was covered in Chapter 6.
(Ans: $ST1_r = 19.405287^h$, $ST1_s = 8.385236^h$)
8. Using α_s and δ_s, compute $ST2_r$ and $ST2_s$ which are also rising and setting times.
(Ans: $ST2_r = 19.632516^h$, $ST2_s = 8.887458^h$)

9. Use the interpolation formula given in equation 8.3.3 to compute the LST for Moonrise.

$$T_r = \frac{12.03 * ST1_r}{12.03 + ST1_r - ST2_r}.$$

(Ans: $T_r = 19.778881^{\mathrm{h}}$)

10. Compute the interpolated setting time using the same formula but substituting in $ST1_s$ and $ST2_s$.

$$T_s = \frac{12.03 * ST1_s}{12.03 + ST1_s - ST2_s}.$$

(Ans: $T_s = 8.750549^{\mathrm{h}}$)

11. Convert the LST times, T_r and T_s, from the previous two steps into their corresponding LMT times for Moonrise and Moonset.
(Ans: $LMT_r = 13.222912^{\mathrm{h}}$, $LMT_s = 2.224687^{\mathrm{h}}$, $GMT_r = 18.222912^{\mathrm{h}}$, $GMT_s = 7.224687^{\mathrm{h}}$)

12. If desired, convert the LMT times to HMS format.
(Ans: $LMT_r = 13^{\mathrm{h}}13^{\mathrm{m}}$, $LMT_s = 2^{\mathrm{h}}13^{\mathrm{m}}$)

8.4 Lunar Distance and Angular Size

The distance from the Earth to the Moon varies throughout the month by as much as 50,000 kilometers. This is true because the Moon's orbit is an ellipse rather than a circle. Moreover, the apparent size of the Moon changes because the distance changes. Calculating these two factors is a very simple task once the Moon's corrected anomaly is known. The distance is given by the equation

$$Dist = \frac{384,401 * (1 - \epsilon^2)}{1 + \epsilon * \cos(C_a + E_c)} \qquad (8.4.1)$$

while the angular diameter is given by

$$\theta = \frac{0.5181 * (1 + \epsilon * \cos(C_a + E_c)}{1 - \epsilon^2}. \qquad (8.4.2)$$

Calculate the distance to the Moon and the Moon's angular diameter for March 12, 1985 at 12^{h} GMT.

1. Calculate C_a and E_c as demonstrated above.
(Ans: $C_a = 40.237641°$, $E_c = 4.062180°$)

2. $T = \cos(C_a + E_c)$.
(Ans: $T = 0.715695$)

3. Compute

$$R = \frac{1 - \epsilon^2}{1 + \epsilon * T}$$

where $\epsilon = 0.054900$ is the eccentricity of the Moon's orbit.
(Ans: $R = 0.959294$)

4. $Dist = 384,401 * R$.
(Ans: $Dist = 369,000$ km)

5. The angular diameter, θ, is

$$\theta = \frac{0.5181}{R}.$$

(Ans: $\theta = 0.540085°$)

6. If desired, convert distance to miles and angular diameter to DMS format.
(Ans: $Dist = 369,000$ km or 229,000 miles, $\theta = 0°32'$)

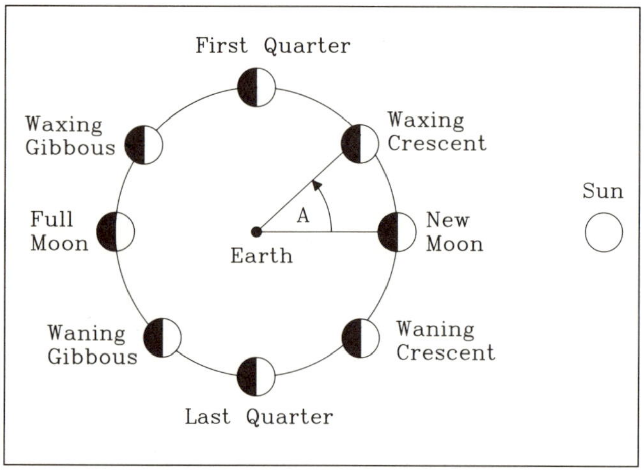

Figure 8.6: Phases of the Moon. *The Moon's phases were an early basis for defining the months. In modern times, the phase cycle defines the synodic month (29.5306 days).*

8.5 Phases of the Moon

As the Moon moves in its orbit around the Earth, the amount of the Moon's surface we see which is illuminated changes. There are three related measures which are sometimes used to quantitatively describe how much of the Moon is illuminated. These three measures are age of the Moon, phase of the Moon, and percent illumination.

Figure 8.6 illustrates the Moon at various points during the month. As can be seen from the figure, the amount of the Moon illuminated as seen from the Earth varies from no illumination at all to full illumination. Notice that the amount of illumination changes when viewed from the Earth but when viewed from outside the Earth, half of the Moon is always illuminated except during an eclipse.

Referring to Figure 8.6, one measure for expressing how much of the Moon is visible is the age of the Moon. In this context, the age of the Moon is an angle which measures how much of its orbit has been completed. This is denoted as $\angle A$ in Figure 8.6 and is measured from the

Moon's current position to the position where the Moon is not illuminated at all (as viewed from the Earth). This position of no illumination is called the New Moon as indicated in the figure. At New Moon, the age of the Moon is 0°. When the age is 90°, half of the visible portion of the Moon is illuminated and it is said to be a First Quarter Moon. At age 180°, the Moon is at maximum illumination and is called a Full Moon. 90° later, when the age is 270°, the Moon is again half illuminated and is said to be a Last Quarter Moon. Notice that First

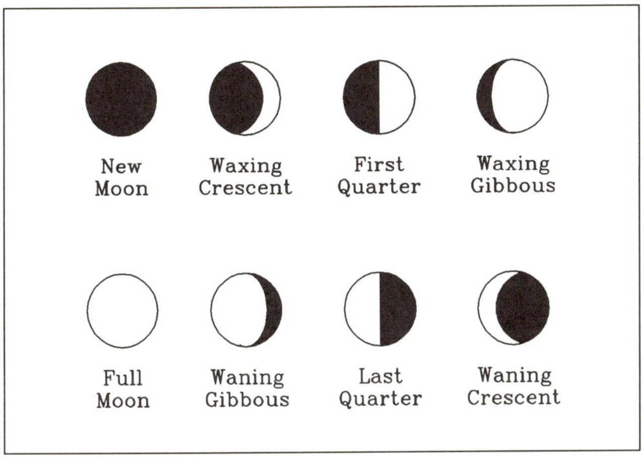

Figure 8.7: *Phases of the Moon as seen from Earth cycles from no illumination (New Moon) to fully illuminated (Full Moon).*

Quarter and Last Quarter Moon are equally illuminated but the portion illuminated is different. Figure 8.7 shows how each of the different phase look from the Earth. The equation for computing the Moon's age (in degrees) is

$$Age = \lambda_t - \lambda_\odot. \tag{8.5.1}$$

Several terms are used to describe the Moon as it orbits the Earth. As the Moon proceeds from New Moon to Full Moon, it is said to be waxing. The age increases from 0° to 180° and the amount of illumination is increasing. When the Moon proceeds from Full Moon back to New Moon, it is said to be waning. During this time, the age increases from 180° to 360° while the amount of illumination is steadily decreasing. When the Moon is at either Full Moon or New Moon, it is on a line connecting the Earth, Moon, and Sun. At Full Moon, the Moon is said to be at the point of opposition with the Sun while it is said to be in conjunction with the Sun at New Moon. Any two objects are at opposition when they are on opposite sides of the Earth and hence have ecliptic longitudes that are 180° apart. Two objects in conjunction are on the same side of the Earth and have the same ecliptic longitude. When at First or Last Quarter, the Moon is said to be in quadrature. Finally, the term gibbous refers to the time when the Moon is between First and Last Quarter. During this time, more than half of the visible portion of the Moon is illuminated as seen from the Earth.

Generally the age of the Moon is expressed in terms of days rather than degrees. In this case, the Moon moves about 12.19° per day so that the age expressed in days is

$$Age = \frac{Age \text{ in Degrees}}{12.19} \tag{8.5.2}$$

or equivalently

$$Age = \frac{\lambda_t - \lambda_\odot}{12.19}. \tag{8.5.3}$$

Using this measure, the New Moon is at 0 days, First Quarter is 7.4 days, Full Moon is 14.8 days, and Last Quarter is 22.1 days.

Let us calculate the age (in degrees and days) of the Moon on January 1, 1985 at 12^h GMT.

1. Compute the Moon's true longitude for the given date.
   (Ans: $\lambda_t = 35.594415°$)
2. Compute the Sun's ecliptic longitude. This value is available as a result of the calculations needed for the first step.
   (Ans: $\lambda_\odot = 281.039905°$)
3. $Age = \lambda_t - \lambda_\odot$.
   (Ans: $Age = -245.445490°$)
4. If necessary, adjust Age to be in the range $[0°, 360°]$.
   (Ans: $Age = 115°$ or 9.4 days)

Perhaps more frequently than age, the amount of illumination is called the phase of the Moon. The phase of the Moon is given by the equation

$$F = \frac{1 - \cos Age}{2} \tag{8.5.4}$$

where Age is measured in degrees. For the previous example, the phase is equal to 0.71. This is closest to a First Quarter Moon.

Notice that when the age is expressed in days, the Moon changes phases (New Moon, First Quarter Moon, Full Moon, Last Quarter Moon) about every 7 days. Thus, the phase of the Moon provides an astronomical event for measuring a week.

Finally, the amount of the Moon illuminated can be expressed as a percentage. The number produced is really the same as the phase but we shall arrive at the answer in a slightly different way. The procedure here for determining how much of the Moon is illuminated is similar to the procedure for determining how much of a planet is illuminated. If I is the Moon's phase angle (calculated below), then the amount of the Moon illuminated is

$$K = \frac{1 + \cos I}{2}. \tag{8.5.5}$$

Again using January 1, 1985 at 12^h GMT, the steps are:

1. Calculate the ecliptic coordinates of the Moon for the given date and the mean anomaly, M_m.
   (Ans: $\lambda_m = 35.667205°$, $\beta_m = -1.718391°$, $M_m = 204.938748°$)
2. Compute the Sun's ecliptic longitude and its mean anomaly, $M_\odot$. Again, these values are available from step 1.
   (Ans: $\lambda_\odot = 281.039905°$, $M_\odot = 358.493854°$)
3. $T = \cos(\lambda_m - \lambda_\odot) * \cos \beta_m$.
   (Ans: $T = -0.416526$)
4. $d = \cos^{-1} T$. The result is the age of the Moon.
   (Ans: $d = 114.615473°$)
5. Compute
   $$R = \frac{1 - 0.0549 * \sin M_m}{1 - 0.0167 * \sin M_m}.$$
   (Ans: $R = 1.022700$)
6. $I = 180° - d - 0.1468 * R * \sin d$. The value I is called the Moon's phase angle.
   (Ans: $I = 65.248038°$)
7. Compute
   $$K = \frac{1 + \cos I}{2}.$$
   (Ans: $K = 0.709345$)
8. Multiply K by 100 to express the illumination as a percentage.
   (Ans: Percent Illuminated = 71%)

Table 8.2 shows the relationship between age, phase, and percent illumination. Each of these measures is simply another way of expressing how much of the Moon is illuminated by the Sun's rays as seen from the Earth.

	Age	Phase	% Illuminated
New Moon	0°	0.0	0
First Quarter	90°	0.5	50
Full Moon	180°	1.0	100
Last Quarter	270°	0.5	50

Table 8.2 Ages, Phase, and Percent Illuminated: *All three measures are different ways of expressing how much of the Moon is visible from the Earth.*

8.6 Eclipses

An eclipse is one of the most exciting astronomical events to observe. In ancient times the sudden disappearance of the Sun or Moon was cause for alarm. In modern times, eclipses are eagerly awaited as opportunities to study the atmosphere surrounding the Sun. This section will describe some of the characteristics of eclipses but determining when and where they occur is beyond the scope of this book. The interested reader can refer to some of the works listed in Appendix C for more information.

There are two types of eclipses: solar and lunar. Both types are caused when the Moon or Earth passes into a shadow, thus blocking out the Sun's light. Figure 8.8 shows the circumstances involving a total lunar eclipse. In this case, the Moon enters the Earth's shadow. The darkest portion of the shadow in Figure 8.8 is called the umbra while the lighter portion of the shadow is called the penumbra. Notice that a total lunar eclipse occurs when the Moon is totally within the umbra. Also notice that a lunar eclipse can be seen from anywhere on the nighttime part of the Earth. When the time of a lunar eclipse is known, compute the Moonrise and Moonset time for an observer. If the time of the eclipse falls within the Moonrise and Moonset times for the observer, then that observer can see the eclipse.

Solar eclipses are caused when the Earth moves into the shadow cast by the Moon. Refer to Figure 8.9. The shadow cast by the Moon is not nearly as long nor as large as that cast by the Earth since the Moon is a smaller object. Therefore, the umbra does not always reach the Earth as shown in Figure 8.9a. This is called an annular solar eclipse and the result as seen from Earth is that the Moon appears as a smaller disk on top of the Sun. Figure 8.9b shows the umbra reaching the Earth. This is called a total solar eclipse. Notice that during a total solar eclipse only a small part of the Earth is in the umbra portion of the Moon's shadow, yet a larger portion of the Earth is in the penumbra. Solar eclipses are thus more difficult to predict since location on the Earth is important.

Eclipses seem to follow a set of rules which can be used to determine if an eclipse will occur. Even though we will not actually compute the occurrence of an eclipse, some of the rules can easily be used to determine whether or not an eclipse is likely. Determining exactly when and where the eclipse will occur is much more difficult.

After a moment's reflection, it should be obvious that eclipses can occur only when the Moon is near opposition to or near conjunction with the Sun. That is, an eclipse can only occur at times near a Full Moon or a New Moon. At a Full Moon, the Earth is between the Sun and Moon so that a lunar eclipse is possible only at a Full Moon. At a New Moon, the Moon is between the Earth and Sun

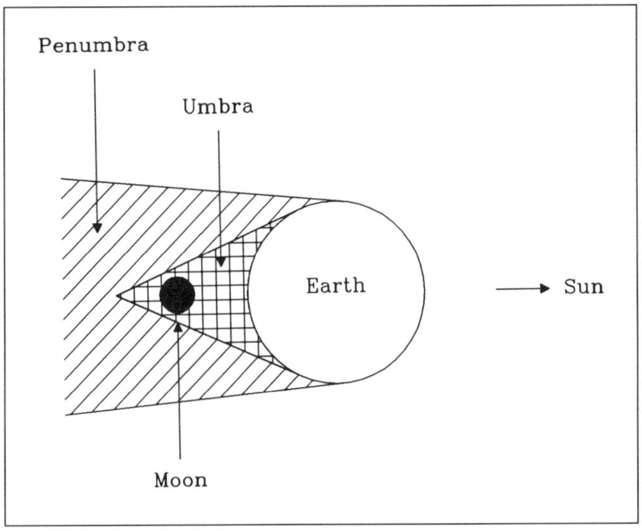

Figure 8.8: Lunar eclipse. *In a lunar eclipse, the Earth passes between the Moon and Sun. Not drawn to scale.*

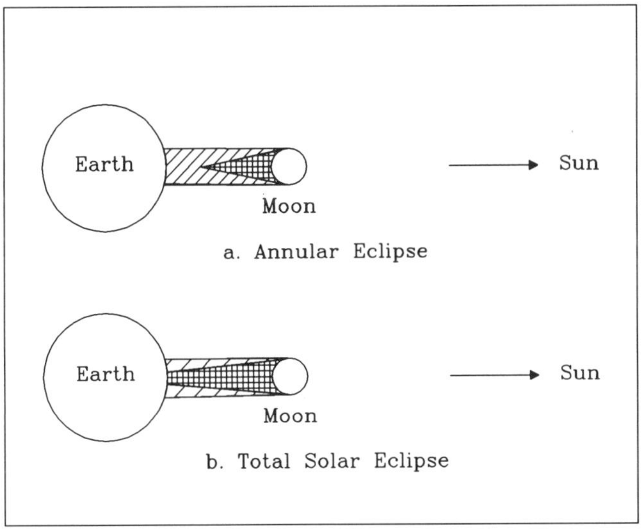

Figure 8.9: Solar eclipse. *In a solar eclipse, the Moon passes between the Earth and Sun. Not drawn to scale.*

so that a solar eclipse is possible only at a New Moon. But eclipses do not occur each month.

Eclipses follow a pattern called the Saros cycle. This cycle is 18 years, 11 days, 8 hours in length but is not exact. There are at least 2 solar eclipses each year, even if only partial, but not more than 5. While at least 2 solar eclipses will occur each year, there may not be any lunar eclipses. If there are, there will be at most 3 lunar eclipses in a single year. However, the total number of eclipses (solar and lunar) for a given year will not exceed 7. A lunar eclipse is often preceded or followed by a solar eclipse which occurrs 2 weeks earlier or later.

During the section for calculating the position of the Moon, the values of the Moon's true orbital longitude and the corrected ecliptic longitude of the ascending node are computed. These are λ_t and Ω' from steps 22 and 23. When the difference between λ_t and Ω' is greater than $12°15'0''$ of $0°$ or $180°$, then a lunar eclipse can not occur. However, if the difference is within $9°30'0''$, then a lunar eclipse must occur assuming that the other rules above check out. Mathematically, if

$$|\lambda_t - \Omega'| > 12°15'0'' \tag{8.6.1}$$

or

$$|\lambda_t - \Omega' - 180°| > 12°15'0'' \tag{8.6.2}$$

a lunar eclipse can not occur. However, if

$$|\lambda_t - \Omega'| < 9°30'0'' \tag{8.6.3}$$

or

$$|\lambda_t - \Omega' - 180°| < 9°30'0'' \tag{8.6.4}$$

a lunar eclipse must occur. The maximum time of a lunar eclipse is 3^h40^m. This includes the time that the Moon is in the Earth's penumbra. The maximum time for the umbral phase of the lunar eclipse is 1^h40^m.

If the difference between λ_t and Ω' is greater than $18°31'0''$ of $0°$ or $180°$, a solar eclipse can not occur. If the difference is within $15°31'0''$ of $0°$ or $180°$, then a solar eclipse must occur. The maximum time that a solar eclipse will last for a given place on the Earth's surface is 12^m24^s for an annular eclipse and 7^m40^s for a total solar eclipse.

In applying the various rules for determining the occurrence of an eclipse, it is important to remember that an eclipse can only occur during Full Moon or New Moon. Thus, there is no point in looking at the values of λ_t and Ω' for a First Quarter Moon.

8.7 About the Program

The program MOON contains the software necessary to implement the algorithms presented in this chapter. As was true in preceding chapters, most of the subroutines are merely renumbered and copied into this program. The subroutine which calculates the position of the Sun and certain other parameters will appear again in the next chapter to calculate the position of the planets. New subroutines included in MOON calculate the position of the Moon, the times of Moonrise and Moonset, the age of the Moon, the distance to the Moon, and the angular diameter of the Moon.

8.8 Sample Problems

1. An observer is at 95°W longitude, 30°N latitude. The date is August 9, 1980 and the observer is in

the Central Standard Time Zone on Daylight Saving Time. If the LMT is 12^h, what are the ecliptic coordinates of the Moon? What are the equatorial coordinates? What are the horizon coordinates?
(Ans: $\lambda_m = 124.536506°$, $\beta_m = -1.397047°$,
$\alpha_m = 8.434972^h$, $\delta_m = 17.772804°$,
$h_m = 75°41'$, $A_m = 146°46'$)

2. For the observer in the last problem, at what time will the Moon rise and set?
(Ans: $LMT_r = 5^h12^m$, $LMT_s = 19^h15^m$)

3. Another observer is located at 30°W longitude, 20°S latitude within the Eastern Standard Time Zone. For the date May 6, 1984, assume that the observer is on Daylight Saving Time. If the LMT is $14^h30^m0^s$, what are the ecliptic coordinates of the Moon? What are the equatorial coordinates? What are the horizon coordinates?
(Ans: $\lambda_m = 128.059516°$, $\beta_m = 4.478048°$,
$\alpha_m = 8.781902^h$, $\delta_m = 22.575935°$,
$h_m = 43°23'$, $A_m = 25°0'$)

4. At what time will the Moon rise and set for the previous observer?
(Ans: $LMT_r = 9^h48^m$, $LMT_s = 21^h5^m$)

5. At 12^h GMT on August 9, 1980, how far away was the Moon? What was its angular diameter?
(Ans: *Dist* = 3.900E05 km or 2.423E05 miles, $\theta = 0°31'$)

6. At $14^h30^m0^s$ GMT on May 6, 1984, how far away was the Moon? What was its angular diameter?
(Ans: *Dist* = 3.680E05 km or 2.287E05 miles, $\theta = 0°32'$)

7. On August 9, 1980 at 12^h GMT, what percentage of the visible portion of the Moon was illuminated?
(Ans: 2%)

8. On May 6, 1984 at $14^h30^m0^s$ GMT, what percentage of the visible portion of the Moon was illuminated?
(Ans: 40%)

9. On August 9, 1980 at 12^h GMT, what was the Moon's age? What was the phase of the Moon?
(Ans: *Age* = 345° (28.3 days), $F = 0.02$ which is closest to a New Moon)

10. On May 6, 1984 at $14^h30^m0^s$ GMT, what was the Moon's age? What was the phase of the Moon?
(Ans: *Age* = 79° (6.5 days), $F = 0.40$ which is closest to a First Quarter Moon)

Chapter 9

The Planets

At last it is time to consider the planets within the solar system. The preceding chapters have led inevitably to this climax by demonstrating how to locate a star, the Sun, and the Moon. With the algorithms presented in this chapter, the reader has a valuable set of mathematical tools for exploring the nighttime sky.

As far as is known today, there are nine major planets in the solar system. Starting from the Sun and proceeding outwards, they are Mercury, Venus, Earth, Mars, Jupiter, Saturn, Uranus, Neptune, and Pluto. In addition, there is a band of asteroids in orbits mainly between Mars and Jupiter. Many have suggested that the asteroid belt is the remnant of an exploded planet. It is speculated that such a planet could have exploded because of impact with another object or due to gravitational pressures generated by the behemoth Jupiter. A more recent theory is that the asteroid belt is the result of thousands of "planetesimals" that subsequently collided and fragmented.

Whether or not there are other planets within the solar system has been a topic of discussion for many years. Searching for a new planet is a difficult task, even with sophisticated telescopes, electronic equipment, and mathematical analysis. Finding a needle in a haystack, to borrow a tired phrase, is almost certainly an easier job. But the search goes on and instant fame is assured to anyone who should stumble across a tenth planet.

Astronomers thought a new planet had been found on March 26, 1859. A Frenchman named Lescarbault claimed discovery of a new planet in an orbit between Mercury and the Sun. He named the planet Vulcan after the god of fire but all attempts to locate this new planet again have failed.

The quest for other planets is not really restricted to the solar system. Astronomers feel quite confident that there must be planets orbiting at least some of the billions of stars in the universe. After all, our Sun appears to be quite ordinary as stars go. Why should our Sun be the only star in the universe with planets? It seems very unlikely but to date, there is no tangible evidence to the contrary.

Exciting as it may be, the discovery of a new planet, within the solar system or otherwise, pales in significance to the discovery of life outside the boundaries of the Earth. The prospect of extraterrestrial life, especially intelligent life, is tantalizing indeed. This prospect has occupied our thoughts in recent years as is evidenced by recent movies, books, and television programs. But is all this just wishful thinking? Are we alone in the cosmos?

Figure 9.1: *These images of the planets in our solar system and the Earth's moon were taken from NASA spacecraft. The picture montage consists of: in the foreground, the Earthrise over the lunar surface with the Sun flare on the edge of the Earth's limb; the first planet above the Moon is Venus; at the top from left to right are the planets Jupiter, Mercury, Mars, and Saturn.*

There is generally great optimism among scientists today, especially astronomers, that life does exist elsewhere. There may even be primitive life forms within our own solar system although the odds look more and more against this as our understanding of the planets increases. But from a practical point of view, unless life is found within the solar system, the question may prove to be unanswerable. At least this is true if the answer is no. In order to prove that life does not exist elsewhere, we will have to explore the entire universe which will take a very long, long time. And during such a massive exploration, will we not find it necessary to colonize other worlds as stopover points? After long eons of time have passed and life is finally found on some distant planet, from whence did it come? Perhaps the new civilization is the remnant of a lost or long forgotten Earth colony ....

Discussion of the planets inevitably leads to speculation on extraterrestrial life. Let us resist further speculation and get back to the main intent of this chapter. To begin, a short description of each of the planets, except Earth, is presented. From there computational algorithms will be presented for locating the planets and for determining some other interesting facts about the planets. Table 9.2 contains the necessary planetary data for use in this chapter.

9.1 Mercury

Mercury is the planet closest to the Sun. It was named after the Roman god, Mercury, who served as a messenger for the other gods. Mercury was also the god of commerce, travel, and thievery. Because of its close proximity to the Sun, Mercury is not often easy to see even though it is visible to the naked eye. It is a small planet with a diameter of about 3,000 miles and a mean distance from the Sun of about 36 million miles. No moons have yet been detected in orbit around Mercury.

Temperatures on the surface soar to around 750° Fahrenheit because Mercury's orbit is so close to the Sun. And yet, temperatures may drop to as low as −300° Fahrenheit on the dark side of the planet. While the density of Mercury is nearly the same as the Earth, its mass is so much smaller (about 0.055 times that of the Earth) that it is doubtful that Mercury's gravitational field is strong enough to hold an atmosphere.

Time as measured on Mercury is substantially different from time measured on Earth. A day on Mercury is about 59 Earth days long while a year is slightly less than 88 Earth days.

9.2 Venus

Venus, sometimes called the Morning or Evening Star, is named after the Roman goddess of love and beauty. When visible, it is one of the brightest objects in the sky

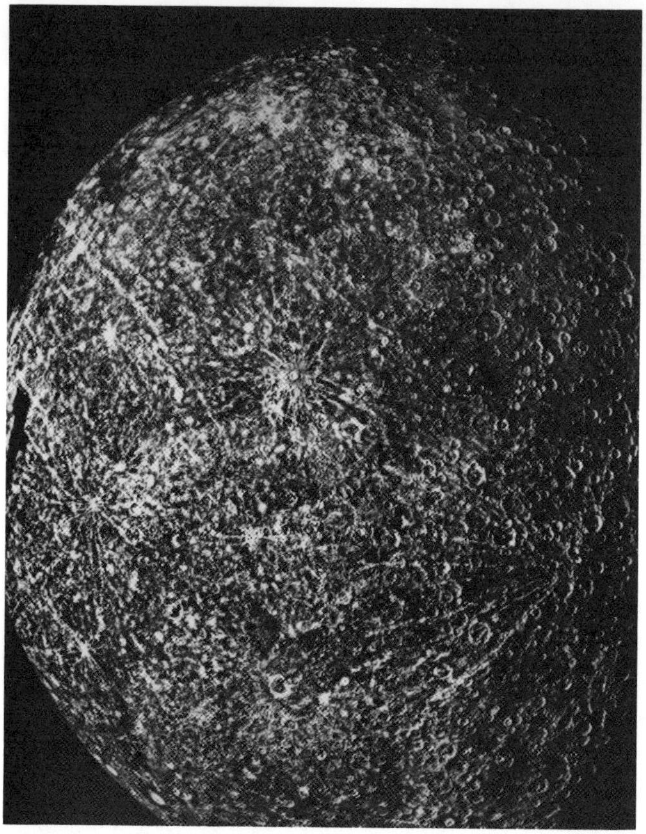

Figure 9.2: *Mercury's surface is pitted with craters caused by the impact of meteors.*

with a magnitude of -4.3. By comparison, the North Star, Polaris, is a second magnitude star. The diameter of Venus is about 7,500 miles making it slightly smaller than the Earth. Its mean distance from the Sun is a little over 67 million miles. There are no known Venusian moons.

Surface temperatures on Venus have been measured to be as high as 900° Fahrenheit. This is hot enough to melt lead so that conditions on Venus are intolerable indeed. In fact, another Roman name, Lucifer, was sometimes used to refer to this Morning Star planet. This may be a more appropriate comparison than comparing Venus to the goddess of love. The extreme temperatures on Venus are most likely due to a greenhouse effect caused by its heavy atmosphere which consists primarily of carbon dioxide.

Venus is unique in the solar system for two reasons. First of all, Venus rotates on its axis from East to West. All other planets in the solar system rotate in the opposite direction. If the Earth rotated from East to West, we would see the Sun rise in the West and set in the East. Secondly, Venus is unique in that its day is longer than its year! Venus rotates on its axis once every 243

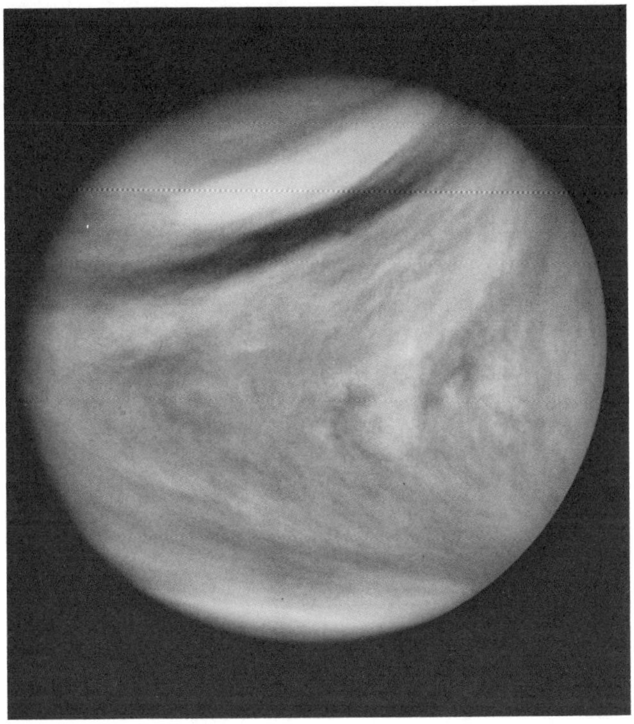

Figure 9.3 *This full-disc NASA picture of Venus was taken by the Pioneer-Venus Orbiter on February 10, 1979. The dense clouds so completely cover the planet that no surface features are visible.*

Earth days while it revolves around the Sun every 225 Earth days.

A manned landing on Venus would be difficult due to the high surface temperature, but unmanned probes can be sent to study the planet. Mariner 2, launched on August 27, 1962, reached Venus on December 14, 1962 to send back scientific data about the planet's surface. Between 1965 and 1978, the Soviets succeeded in landing 10 space probes on Venus. Pictures sent back show a bleak and barren landscape.

9.3 Mars

Perhaps no planet has stirred as much controversy as Mars, the red planet. Named for the Roman god of war, Mars has been scrutinized for signs of life ever since the 1870s when an Italian astronomer, Giovanni V. Schiaparelli, reported seeing canals on the planet's surface. Science fiction writers have capitalized on this and written stories about invading "little green men from Mars." The classic example of this is the October 30, 1938 radio broadcast of the *War of the Worlds.* This story of an imaginary Martian invasion wreaked havoc as people panicked in fear.

Mariner 9 reached Mars in 1971 but found no signs of Martian canals, built by "little green men" or otherwise. Likewise, the Viking space probe which landed in 1976 found no evidence of Martian canals or life, intelligent or otherwise. The canals reportedly seen by Schiaparelli and others may well be the result of dust storms on the surface of Mars.

Figure 9.4: *The four largest Martian volcanoes are portrayed in this NASA picture of Mars taken by the Viking 1 Orbiter from a range of 350,000 miles. The topmost is Olympus Mons, its mountain reaching a height of about 15 miles with the base 300 miles across. The middle volcano in the row of three straddles the equator. The great rift, stretching some 2,200 miles can be seen faintly between the terminator and the bright limb in the lower half of the disc. The bright area at the bottom is Argyre, a large impact basin.*

Both the Mariner 9 and Viking space probes sent back a massive amount of data to the Earth. The Martian atmosphere is chiefly carbon dioxide. Surface temperatures are somewhat colder than on Earth with temperatures that may be as low as −120° Fahrenheit and as high as about 80° Fahrenheit. A surprise to astronomers, the space probes found that the surface of Mars is pitted with craters which are partially filled in with wind blown sand. In addition, a gigantic volcano was found on Mars which is larger than any known upon the Earth. And if that is not impressive enough, a Martian canyon was photographed which is far deeper than the Grand Canyon.

Mars, like the Earth, has seasons. The polar caps grow and shrink in much the same way as the Earth's polar regions grow and shrink with the changing seasons. This, along with the dust storms, may account for apparent changes on the surface of Mars.

Mars has a mean distance from the Sun of about 142 million miles. Its day is 24^h37^m which is a little longer than the Earth's day. A Martian year is about 687 Earth days making it a little less than twice an Earth year.

Two Martian moons have been found. These are named Phobos and Deimos and were discovered in 1877. Phobos, photographed from the Mariner space probe, revolves about Mars in 7.5^h. Thus, Phobos revolves about Mars faster than Mars itself revolves on its own axis. Phobos is the only known moon in the solar system which revolves about the planet faster than the planet itself revolves.

9.4 Asteroids

Astronomers had long suspected that there must be a planet between Mars and Jupiter. The gap between the two planets just seemed too big. Finally, an Italian astronomer named Giuseppe Piazzi discovered a "planet" in 1801. This new object was named Ceres and was found to have a diameter of about 480 miles. Astronomers soon discovered, however, that there are actually several large objects in an orbit between Mars and Jupiter. Between 1802 and 1807, three more objects were found: Pallas with a diameter of 304 miles, Vesta with a diameter of 240 miles, and Juno with a diameter of 120 miles. Since then, a total of about 30,000 objects have been identified as belonging to the asteroid belt of which about 4000 have been formally cataloged and named. Some have recently estimated the number of asteroids to be as high as 500,000.

Most of the asteroids that have been discovered are far smaller than Ceres, averaging a mere 50 miles in diameter. Vesta, the brightest known asteroid, is not as large as Ceres but it can sometimes be seen with the naked eye. Table 9.1 lists several asteroids, their size, and relative distance from the Sun.

It might appear at first glance that the asteroid belt poses a serious threat to space travel. Most attempts to visualize the asteroid belt depict a zone of densely packed boulders that barely leave enough room for a hapless spaceship to pass through. But this view is most likely wrong. Isaac Asimov estimates that the asteroids, except for occasional clusters, may have an average distance between them of some 10 million miles. If true, the probability of encountering an asteroid, much less colliding with one, is small. The Voyager space probe passed through the asteroid belt on its way to Jupiter and Saturn without any mishap whatsoever which serves as further evidence that little danger is presented by the asteroid belt.

Asteroid	Diameter (km)	Distance (A.U.)
Ceres	1003	2.767
Davida	318	3.182
Eros	14	1.458
Eunomia	248	2.645
Geographos	3	1.244
Hebe	220	2.426
Hygiea	414	3.151
Icarus	1.4	1.077
Iris	211	2.385
Juno	247	2.670
Nemausa	80	2.366
Pallas	608	2.767
Psyche	249	2.923
Vesta	528	2.361

Table 9.1: *Typical Asteroids*

The initial search for a "missing" planet centered on the region between Mars and Jupiter. However, not all of the asteroids have orbits which lie within this region of the solar system. Many asteroids have orbits that carry them inside the orbit of Mars and some even have orbits that carry them inside the orbit of Earth. The asteroids with orbits at Jupiter's two Lagrangian points are known as the Trojan asteroids and account for about ninety-five percent of all known asteroids. The Amor group of asteroids travel inside the orbit of Mars while the Apollo group travel inside the orbit of Earth.

In addition to classifying asteroids based upon their orbital trajectory, astronomers also classify asteroids by their spectral reflectivity. Originally only two classes where used: C (carbonaceous) and S (stony). This classification system was recently increased to a total of 14 classes including M (metallic), F (flat), and R (red).

The sheer number of asteroids makes it virtually impossible to assign each the name of a mythological god, person, mountain, or some otherwise "interesting" name. Instead, astronomers have adopted a provisional naming scheme for the asteroids until their orbit is accurately determined. Once the asteroid's orbit is accurately known, a definitive number is assigned. The naming scheme adopted consits of a number followed by two letters followed by an optional number. The first number is the year in which the asteroid was discovered while the first letter indicates the month in which the asteroid was discovered. But, instead of using A for January, B for February, and so on, astronomers use the letter A for

January 1-15, B for January 16-31, C for February 1-15, D for February 16-29, and so forth. The second letter in the naming convention is used in case more than one asteroid is discovered within the same time period with A representing the first one discovered, B the second, and so on. The optional number previously mentioned is just in case more than 26 asteroids are discovered within a time period covered by a single letter. For example, asteriod 1984 BC was the third (letter C) asteroid discovered between January 16 and January 31 (letter B) in 1984.

9.5 Jupiter

Jupiter, by far the largest of the planets, is appropriately named for the ruler of the gods. It has a diameter of over 88,000 miles making its diameter about 11 times larger than the Earth. Along with Saturn, Uranus, and Neptune, Jupiter is one of the so called gas giants. Its atmosphere is mostly hydrogen with some helium, methane, and ammonia. The temperature of the cloud tops of Jupiter is estimated to be as low as -228° Fahrenheit. Hardly a hospitable place for humans!

Galileo was the first to observe Jupiter through a telescope. In 1610 he discovered four moons orbiting about the planet which Simon Mayr later named Io, Europa, Ganymede, and Callisto. Twelve more moons have been discovered since then for a total of sixteen (in 1988)!

When passed by the Voyager 1 space probe, several interesting facts were discovered about Jupiter's moons. Callisto is heavily cratered and may be the most cratered object in the solar system. By contrast, Europa is incredibly smooth, so smooth that Europa has been compared to a billiard ball. Perhaps Europa, like Ganymede, is ice covered. Io is unlike any of the other moons of Jupiter and unlike any other known moon in the solar system for that matter. About eight active volcanoes have been discovered on the surface of Io. But most exciting of all, Voyager 1 detected the presence of a ring around Jupiter, a ring too small to be seen from the Earth.

The most prominent feature of Jupiter is the Great Red Spot. From data sent back in 1973 and 1974 by the Pioneer space probes, the Red Spot has been measured to be 25,000 miles long and 8,000 miles wide. The Earth would easily fit into the Red Spot with room to spare. But the Red Spot appears to be shrinking and may disappear in a century or so.

Jupiter's atmosphere also contains alternating light zones and dark belts of clouds. These belts and zones are bands of high and low pressure parallel to its equator and are made parallel by Jupiter's rapid rotation. Jupiter rotates on its axis in less than 10 hours so that objects at its equator are whirling around at nearly 27,000 miles per hour. By comparison, objects at the Earth's equator rotate at a mere 1,000 miles per hour.

Figure 9.5: *Jupiter and its four planet-size moons, called the Galilean satellites, were photographed by Voyager 1 and assembled into this NASA collage. They are not to scale but are in their relative positions. Io (upper left) is nearest Jupiter, then Europa (center); Ganymede and Callisto (lower right). Twelve other much smaller satellites circle Jupiter, four inside Io's orbit and the others millions of miles from the planet.*

9.6 Saturn

Saturn was the most distant planet known to the ancients. Lying about 890 million miles from the Sun, the ringed planet is the most beautiful in our solar system. It is second in size only to Jupiter and has a diameter of 74,000 miles. Like Jupiter, Saturn's atmosphere appears to be mostly hydrogen. But Saturn's density is less than Jupiter's density and is so low that Saturn would float in water. That is, if a large enough ocean could be found!

Saturn has at least 17 moons, one of which was discovered by the Pioneer 11 space probe in 1979. Titan, the largest of Saturn's moons, has an atmosphere. It is the only known satellite in the solar system to have an atmosphere. Some believe that the rings which encircle Saturn are the remnants of a Saturnian moon. The rings of Saturn were discovered soon after the telescope was invented. In 1655, the Dutch astronomer Christian Huyghens explained why they seem to appear and then disappear. Twenty years later, Giovanni

Figure 9.6: *This montage of NASA images of the Saturnian system was prepared from images taken by the Voyager 1 spacecraft during its Saturn encounter in November 1980. This artist's view shows Dione in the forefront, Saturn rising behind, Tethys and Mimas fading in the distance to the right, Enceladus and Rhea off Saturn's rings to the left, and Titan in its distant orbit at the top.*

Domenico Cassini noticed that a dark gap separated the ring into two parts. This gap is called the Cassini division in his honor. However, the Voyager 1 and Voyager 2 spacecraft clearly demonstrated that the rings are far more numerous and complex than imagined. One ring is even braided!

From the Voyager flybys in November 1980 and August 1981, our knowledge of Saturn has greatly expanded. But for all our efforts, our understanding of this planet is still very primitive. Saturn remains as mysterious as ever and still beckons us like a beautiful siren.

9.7 Uranus

Uranus was discovered by Sir William Herschel on March 13, 1781. The discovery catapulted Herschel to fame and assured him a place in history. But others had seen Uranus without realizing it was a planet. Just barely visible to the naked eye, the British Astronomer Royal John Flamsteed included it as a star on a star chart produced in 1690.

Uranus is one of the gas giants. It has a diameter of 29,000 miles and lies about 1.8 billion miles from the Sun. Conditions on Uranus are inhospitable with an atomosphere consisting of methane and hydrogen. The cloud

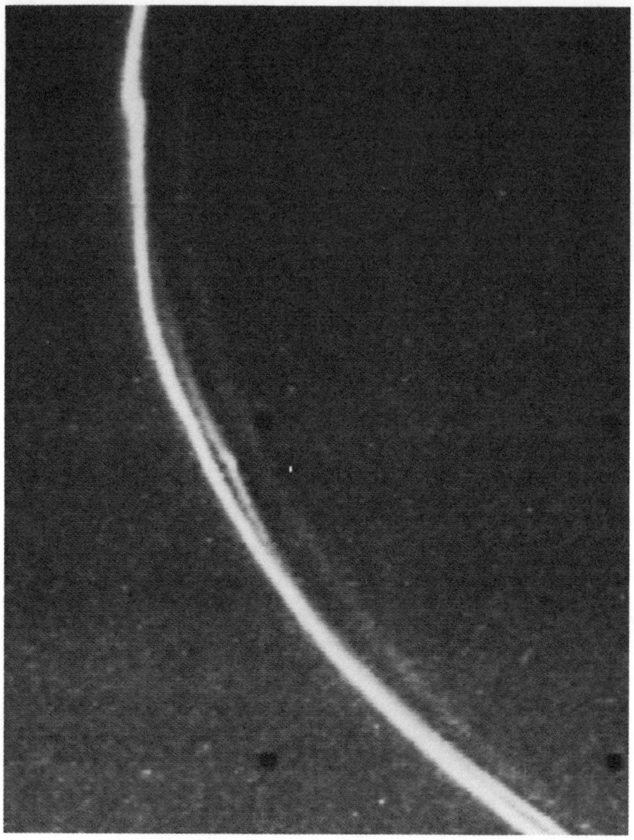

Figure 9.7: *Saturn's F, or outermost ring was photographed from the unilluminated face of the rings by Voyager 1 at a range of 470,000 miles. Complex structure is evident, with several components seen. Two narrow, braided, bright rings that trace distinct orbits are evident. Visible is a broader, very diffuse component about 20 miles in width. Also seen are "knots," which probably are local clumps of ring material, but may be mini-moons.*

tops have a temperature of about −280° Fahrenheit. In 1977 it was discovered that rings exist around Uranus and in January 1986, the Voyager 2 flyby confirmed the existence of fifteen moons around the planet.

An interesting feature of Uranus is that it is "lying down" in its orbit. That is, its axis of rotation is almost in its orbital plane. The axis of rotation for all the other planets in the solar system is more nearly perpendicular to the planet's orbital plane.

9.8 Neptune

Before the discovery of Neptune, the only hope for discovering a new planet was more or less by chance. No one pointed their telescope to a spot in the sky and expected to find a new planet there. But two astronomers

Figure 9.8: *Voyager 2 is seen on its approach to the planet Uranus, which in this NASA computer-generated view is only 2 hours 30 minutes away. Voyager passed within 50,600 miles of Uranus' cloudtops on January 24, 1986.*

independently and at about the same time calculated the position of where Neptune should be. The two astronomers were John C. Adams of England and Urbain J. J. Leverrier of France. Johann G. Galle of Germany sighted the planet where predicted on September 23, 1846.

Very little is known about Neptune since it is so far away. Its mean distance from the Sun is 2.8 billion miles. With a diameter of 30,000 miles, Neptune is almost as big as Uranus. The atmosphere of Neptune is probably similar to Uranus consisting of hydrogen, methane, and helium.

In 1982, astronomer Edward Guinan announced the tentative discovery of a ring encircling Neptune. Two observations made in 1984 at two separate observatories provide further evidence that Neptune may indeed have a ring. If further observations confirm Guinan's discovery, then within the last two decades the number of confirmed ringed planets has grown from 1 (Saturn) to 4 (Saturn, Jupiter, Uranus, and Neptune).

Two moons have been discovered orbiting Neptune. The largest, Triton, is larger than the Moon. Triton has an unstable orbit around Neptune that is "rapidly" decaying. In 1966, astronomer Thomas McCord predicted that Triton will collide with Neptune in about 100 million years. The result may be another asteroid belt in the outer reaches of our solar system, or the creation of a ringed planet that will rival Saturn.

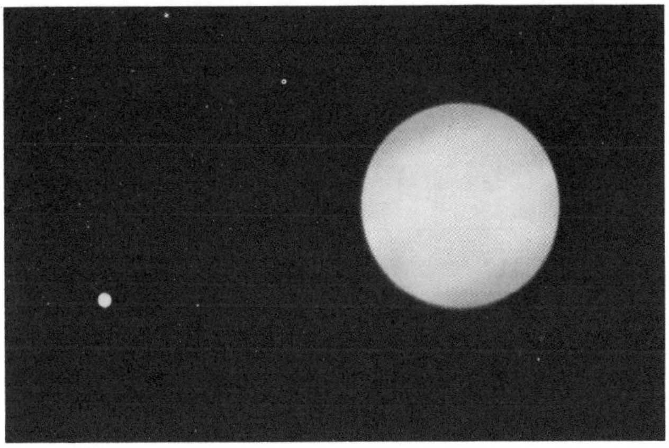

Figure 9.9: *Neptune and its largest moon, Triton, as visualized by a NASA artist.*

Figure 9.10: *Pluto and its moon, Charon, as depicted by a NASA artist.*

9.9 Pluto

While little is known about Neptune, even less is known about Pluto. Even with the largest telescopes, Pluto appears only as a point of light so that no surface features can be seen from the Earth. Discovered in 1930 by the American astronomer Clyde W. Tombaugh, Pluto is about 3.7 billion miles away from the Sun. Its diameter is estimated to be 1,400 miles making it much smaller than Mercury. Surface temperatures are esti-

Table 9.2
Planetary Data—Part 1

	Sun	Mercury	Venus	Earth	Mars	Jupiter
T_p, Period (in tropical years)		0.24085	0.61521	1.00004	1.88089	11.86224
ε_p, Longitude at epoch 1980.0	278.833540	231.2973	355.73352	98.833540	126.30783	146.966365
ϖ_p, Longitude of perihelion	282.596403	77.1442128	131.2895792	102.596403	335.6908166	14.0095493
ϵ, Eccentricity	0.016718	0.2056306	0.0067826	0.016718	0.0933865	0.0484658
A, Semi-major axis (in AU)	1	0.3870986	0.7233316	1	1.5236883	5.202561
ι, Inclination of Orbital Plane	0	7.0043579	3.394435	0	1.8498011	1.3041819
Ω_p, Longitude of the ascending node		48.0941733	76.4997524		49.4032001	100.2520175
θ_0, Angular size at 1 AU (arcsec)	1,919.26	6.74	16.92		9.36	196.74
Mass (relative to Earth)	332946	0.055274	0.815005	1	0.107447	317.833
Radius (in km)	696000	2439	6052	6378	3394	71400
Length of Day (relative to Earth)		58.646	243.017	1	1.0260	0.4101

mated to be in the neighborhood of −300° Fahrenheit. Pluto has one moon, named Charon, discovered in 1978 by James Christy and Robert Harrington of the United States Naval Observatory.

Due to the relatively small size of Pluto, some astronomers have suggested that Pluto might once have been one of Neptune's moons. It does seem strange that so small a planet should be out there among so many gigantic planets. Pluto's orbit overlaps Neptune's orbit which lends some credence to the "escaped moon" theory. Other astronomers believe that Pluto might be a captured comet or space debris. Until a space probe manages to relay back data from a much closer range, about all we can do is speculate about this small planet.

9.10 Locating the Planets

Calculating the location of a planet is by far the most ambitious calculation in this book. As in the case of locating the Sun or Moon, the main idea is to be able to calculate the true anomaly from which it can be determined how far the planet has moved along its orbit. However, the process is complicated depending upon whether or not the planet is an inner or outer planet and because the position of the Earth must be taken into consideration. (The inner planets are Mercury and Venus while the outer planets are Mars, Jupiter, Saturn, Uranus, Neptune, and Pluto.)

Calculating the position of a planet will require computing the planet's position with respect to the Sun just as was done in calculating the position of the Sun in Chapter 7. But this computation gives the planet's heliocentric coordinates which is the location of the planet with respect to the Sun. What is really desired is the planet's geocentric coordinates which is the location of the planet with respect to the Earth. Hence, it is necessary to compute the location of the Earth as well as the location of the planet.

During the calculation of a planet's position, it is often important to distinguish between a calculation for a planet and a calculation for Earth. To avoid confusion, the variables with subscript e will refer to the Earth. Variables with subscript p will refer to the desired planet.

The overall steps required are as follows:

- Compute the planet's position with respect to the Sun. This will give M_p, the mean anomaly of the planet.
- Compute the planet's heliocentric longitude L_p and latitude Λ_p.
- Compute the planet's true anomaly v_p.
- Compute the Earth's position in the ecliptic plane. This gives the Earth's mean anomaly M_e.
- Compute the Earth's heliocentric longitude L_e.
- Compute the Earth's true anomaly v_e.
- Given the Earth's heliocentric location and the planet's heliocentric location, project the planet's

Table 9.2 Planetary Data—Part 2					
	Saturn	Uranus	Neptune	Pluto	Moon
T_p, Period (in tropical years)	29.45771	84.01247	164.79558	247	
ε_p, Longitude at epoch 1980.0	165.322242	228.0708551	260.3578998	209.439	64.975464
ϖ_p, Longitude of perihelion	92.6653974	172.7363288	47.86272148	222.972	349.383063
ϵ, Eccentricity	0.0556155	0.0463232	0.0090021	0.25387	0.054900
A, Semi-major axis (in AU)	9.554747	19.21814	30.10957	39.78459	
ι, Inclination of Orbital Plane	2.4893741	0.7729895	1.7716017	17.137	5.145396
Ω_p, Longitude of the ascending node	113.4888341	73.8768642	131.5606494	109.941	151.950429
θ_0, Angular size at 1 AU (arcsec)	165.60	65.8	62.2	8.2	
Mass (relative to Earth)	95.159	14.5	17.204	0.0026?	0.01228
Radius (in km)	68000	25600	24300	1150	1738
Length of Day (relative to Earth)	0.4440	0.718	0.768	6.3867	

location onto the ecliptic plane to obtain its geocentric location.

The process of projecting the planet's heliocentric coordinates onto the ecliptic plane depends upon whether the planet is an inner or outer planet. Both will be demonstrated below.

Several equations are required for this rather lengthy process. The planet's mean anomaly is given by

$$M_p = \frac{360 * D}{365.2422 * T_p} + \varepsilon_p - \varpi_p \qquad (9.10.1)$$

where T_p is the planet's period in tropical years obtained from Table 9.2, D is the number of days since the standard epoch 1980.0, ε_p is the planet's ecliptic longitude at the standard epoch, and ϖ_p is the ecliptic longitude of the planet at the moment of perihelion. (The planet's period is how many tropical years are required to complete one orbit around the Sun.) Notice how similar this equation is to that required for computing the Sun's mean anomaly $M_{\odot}$.

The planet's heliocentric longitude is

$$L_p = \frac{360 * D}{365.2422 * T_p} + \varepsilon_p + \left(\frac{360 * \epsilon_p}{\pi}\right) * \sin M_p \quad (9.10.2)$$

where ϵ_p is the planet's orbital eccentricity. The heliocentric latitude is

$$\sin \Lambda_p = \sin(L_p - \Omega_p) * \sin \iota_p \qquad (9.10.3)$$

where Ω_p is the planet's ecliptic longitude of the ascending node and ι_p is the planet's orbital inclination. Given L_p, the planet's true anomaly is

$$\upsilon_p = L_p - \varpi_p \qquad (9.10.4)$$

or equivalently

$$\upsilon_p = M_p + \left(\frac{360 * \epsilon_p}{\pi}\right) * \sin M_p. \qquad (9.10.5)$$

As in the case of the Sun, this is an approximation which avoids the need for an iterative solution to Kepler's equation.

Once the planet's coordinates are computed, the same computations are required for the Earth. That is, equations 9.10.1, 9.10.2, and 9.10.4 are computed using values for the Earth instead of the planet. Equation 9.10.3 is not required because the Earth's heliocentric latitude is 0° since the Earth's orbit lies completely within the ecliptic plane. This is easily seen by substituting $\iota_e = 0°$ into equation 9.10.3.

When the coordinates for the planet and Earth are completed, the distance from the Sun to the Earth and the distance from the planet to the Sun are required. The distance from the Sun in A.U. is given by

$$R = \frac{A * (1 - \epsilon^2)}{1 + \epsilon * \cos \upsilon} \qquad (9.10.6)$$

where A is the length (measured in A.U.) of the orbital

semi-major axis. For the Earth, A is 1 while Table 9.2 gives a value for each of the planets.

The last equations required project the heliocentric coordinates onto the ecliptic plane to produce geocentric ecliptic coordinates. Rather than repeating these lengthy equations, refer to steps 21-36 below.

This lengthy process will be demonstrated by two examples. Let us calculate the location of Venus and Saturn for December 30, 1984 at 22^{h} LMT. Assume the observer is not on Daylight Saving Time but is in the Eastern Standard Time Zone at 78°W longitude, 38°N latitude. Refer to Table 9.2 for various planetary orbital data.

9.10.1 Calculations for an Inner Planet

1. Convert the LMT to GMT, GST, and LST times and adjust the date if necessary.
(Ans: $GMT = 3.0^{\mathrm{h}}$, $GST = 9.648580^{\mathrm{h}}$,
$LST = 4.448580^{\mathrm{h}}$, Date $= 12/31/1984$)
2. Using Julian dates, compute the number of days since January 0.0 for the year given.
(Ans: $Days = 366$)
3. Compute T, the number of leap years since 1980.
(Ans: $T = 2$)
4. Compute $D = T + 365 * (Year - 1980) + Days$.
(Ans: $D = 1,828$)

Steps 5-11 calculate the planet's true anomaly v_p and radius vector R_p.

5. Compute
$$N_p = \frac{360 * D}{365.2422 * T_p}$$
where T_p is the planet's period in tropical years obtained from Table 9.2.
(Ans: $T_p = 0.61521$ for Venus,
$N_p = 2,928.696418°$)
6. If necessary, convert N_p to the range $[0°, 360°]$.
(Ans: $N_p = 48.696418°$)
7. $M_p = N_p + \varepsilon_p - \varpi_p$ where M_p is the planet's mean anomaly, ε_p is the ecliptic longitude at epoch 1980, and ϖ_p is the ecliptic longitude at perihelion.
(Ans: $\varepsilon_p = 355.73352°$, $\varpi_p = 131.2895792°$,
$M_p = 273.140358°$)
8. Compute the planet's heliocentric longitude,
$$L_p = N_p + \varepsilon_p + \left(\frac{360 * \epsilon_p}{\pi}\right) * \sin M_p$$
where ϵ_p is the planet's orbital eccentricity.
(Ans: $\epsilon_p = 0.0067826$, $L_p = 403.653876°$)
9. Convert L_p to the range $[0°, 360°]$ if necessary.
(Ans: $L_p = 43.653876°$)
10. $v_p = L_p - \varpi_p$ where v_p is the planet's true anomaly.
(Ans: $v_p = -87.635703°$)
11. Compute the planet's radius vector length
$$R_p = \frac{A * (1 - \epsilon_p^2)}{1 + \epsilon_p * \cos v_p}.$$
R_p is the distance from the Sun to the planet expressed in AU units.
(Ans: $A = 0.7233316$ AU, $R_p = 0.723096$ AU)

Steps 12-18 are identical to 5-11 except that the calculations are done on the Earth to produce the Earth's true anomaly v_e and radius vector R_e.

12. Compute
$$N_e = \frac{360 * D}{365.2422 * T_e}$$
where T_e is the Earth's period in tropical years.
(Ans: $T_e = 1.00004$, $N_e = 1,801.691255°$)
13. Convert N_e to the range $[0°, 360°]$.
(Ans: $N_e = 1.691255°$)
14. $M_e = N_e + \varepsilon_e - \varpi_e$ where M_e is the Earth's mean anomaly, ε_e is the ecliptic longitude for Earth at epoch 1980, and ϖ_e is the Earth's ecliptic longitude at perihelion.
(Ans: $\varepsilon_e = 98.83354°$, $\varpi_e = 102.596403°$,
$M_e = -2.071608°$)
15. Compute the Earth's heliocentric longitude
$$L_e = N_e + \varepsilon_e + \left(\frac{360 * \epsilon_e}{\pi}\right) * \sin M_e.$$
(Ans: $\epsilon_e = 0.016718$, $L_e = 100.455544°$)
16. Convert L_e to the range $[0°, 360°]$.
(Ans: no change in L_e)
17. Compute $v_e = L_e - \varpi_e$ where v_e is the Earth's true anomaly.
(Ans: $v_e = -2.140859°$)
18. Compute the Earth's radius vector
$$R_e = \frac{1 - \epsilon_e^2}{1 + \epsilon_e * \cos v_e}$$
where R_e is the distance from the Sun to the Earth expressed in AU units.
(Ans: $R_e = 0.983293$ AU)

19. $T = \sin(L_p - \Omega_p) * \sin \iota_p$ where Ω_p is the planet's ecliptic longitude of the ascending node and ι_p is the planet's orbital inclination.
(Ans: $\Omega_p = 76.4997524°$, $\iota_p = 3.394435°$, $T = -0.032114$)

20. $\Lambda_p = \sin^{-1} T$ where Λ_p is the planet's heliocentric latitude.
(Ans: $\Lambda_p = -1.840317°$)

21. $y = \sin(L_p - \Omega_p) * \cos \iota_p$.
(Ans: $y = -0.541430$)

22. $x = \cos(L_p - \Omega_p)$.
(Ans: $x = 0.840133$)

23. Compute
$$t = \tan^{-1}\left(\frac{y}{x}\right).$$
(Ans: $t = -32.800051°$)

24. Add a quadrant adjustment factor to t to remove the ambiguity of the inverse tangent function. This is necessary since t will be used for calculating a longitude which must be in the range $[0°, 360°]$.
(Ans: $Adjustment = 360°$, $t = 327.199949°$)

25. $L' = \Omega_p + t$.
(Ans: $L' = 403.699702°$)

26. $R' = R_p * \cos \Lambda_p$.
(Ans: $R' = 0.722723°$)

27. $z = L_e - L'$.
(Ans: $z = -303.244158°$)

Steps 28-32 are for an inner planet only. These steps will be different for an outer planet.

28. $y = R' * \sin z$.
(Ans: $y = 0.604443$)

29. $x = R_e - R' * \cos z$.
(Ans: $x = 0.587090$)

30. Compute
$$B = \tan^{-1}\left(\frac{y}{x}\right).$$
(Ans: $B = 45.834376°$)

31. Adjust B to account for the ambiguity of the inverse tangent function. This adjustment is necessary since B will be used to calculate the ecliptic longitude which must be in the range $[0°, 360°]$.
(Ans: No adjustment needed.)

32. $\lambda_p = 180° + L_e + B$.
(Ans: $\lambda_p = 326.289920°$)

33. If necessary, adjust λ_p to the range $[0°, 360°]$.
(Ans: no adjustment)

34. $y = R' * \tan \Lambda_p * \sin(\lambda_p - L')$.
(Ans: $y = 0.022663$)

35. $x = R_e * \sin(-z)$.
(Ans: $x = -0.822369$)

36. Compute
$$\beta_p = \tan^{-1}\left(\frac{y}{x}\right).$$
Note that no quadrant adjustment is needed here since the inverse tangent returns a value in the proper range. In fact, computing a quadrant adjustment will give an incorrect answer!
(Ans: $\beta_p = -1.578580°$)

37. Convert the ecliptic coordinates λ_p and β_p to their corresponding equatorial coordinates.
(Ans: $\alpha_p = 21^h56^m$, $\delta_p = -14°14'$)

38. Finally, convert the equatorial coordinates to the appropriate horizon coordinates.
(Ans: $h_p = -14°40'$, $A_p = 263°12'$)

Needless to say, this is a long process indeed. The algorithm for locating one of the outer planets is not any shorter. For Saturn on the given date and at the given location, the steps are as follows:

9.10.2 Calculations for an Outer Planet

1. Convert LMT to GMT, GST, and LST and adjust the date if needed.
(Ans: $GMT = 3^h$, $GST = 9.648580^h$, $LST = 4.448580^h$, Date = 12/31/1984)

2. Use Julian dates to compute the number of days since January 0.0 for the year given.
(Ans: $Days = 366$)

3. Compute T, the number of leap years since 1980.
(Ans: $T = 2$)

4. $D = T + 365 * (Year - 1980) + Days$.
(Ans: $D = 1828$)

5. Compute
$$N_p = \frac{360 * D}{365.2422 * T_p}$$
where T_p is the planet's period obtained from Table 9.2.
(Ans: $T_p = 29.45771$ for Saturn, $N_p = 61.164406°$)

6. If necessary, convert N_p to the range $[0°, 360°]$.
(Ans: $N_p = 61.164406°$)

7. $M_p = N_p + \varepsilon_p - \varpi_p$ where M_p is the planet's mean anomaly, ε_p is the ecliptic longitude at epoch 1980, and ϖ_p is the ecliptic longitude at perihelion.
(Ans: $\varepsilon_p = 165.322242°$, $\varpi_p = 92.6653974°$, $M_p = 133.821500°$)

8. Compute the planet's heliocentric longitude
$$L_p = N_p + \varepsilon_p + \left(\frac{360 * \epsilon_p}{\pi}\right) * \sin M_p$$
where ϵ_p is the planet's orbital eccentricity.
(Ans: $\epsilon_p = 0.0556155$, $L_p = 231.084836°$)
9. Convert L_p to the range $[0°, 360°]$ if necessary.
(Ans: No adjustment needed)
10. $\upsilon_p = L_p - \varpi_p$ where υ_p is the planet's true anomaly.
(Ans: $\upsilon_p = 138.419439°$)
11. Compute the planet's radius vector length
$$R_p = \frac{A * (1 - \epsilon_p^2)}{1 + \epsilon_p * \cos \upsilon_p}$$
where A is the planet's semi-major axis expressed in AU units.
(Ans: $A = 9.554747$ AU, $R_p = 9.938658$ AU)
12. Compute
$$N_e = \frac{360 * D}{365.2422 * T_e}$$
where T_e is the Earth's period.
(Ans: $T_e = 1.00004$, $N_e = 1,801.691255°$)
13. Convert N_e to the range $[0°, 360°]$.
(Ans: $N_e = 1.691255°$)
14. $M_e = N_e + \varepsilon_e - \varpi_e$ where M_e is the Earth's mean anomaly, ε_e is the ecliptic longitude for Earth at epoch 1980, and ϖ_e is the Earth's ecliptic longitude at perihelion.
(Ans: $\varepsilon_e = 98.83354°$, $\varpi_e = 102.596403°$, $M_e = -2.071608°$)
15. Compute the Earth's heliocentric longitude
$$L_e = N_e + \varepsilon_e + \left(\frac{360 * \epsilon_e}{\pi}\right) * \sin M_e$$
where ϵ_e is the Earth's orbital eccentricity.
(Ans: $\epsilon_e = 0.016718$, $L_e = 100.455544°$)
16. Convert L_e to the range $[0°, 360°]$.
(Ans: no change in L_e)
17. $\upsilon_e = L_e - \varpi_e$ where υ_e is the Earth's true anomaly.
(Ans: $\upsilon_e = -2.140859°$)
18. Compute the Earth's radius vector length
$$R_e = \frac{1 - \epsilon_e^2}{1 + \epsilon_e * \cos \upsilon_e}.$$
(Ans: $R_e = 0.983293$ AU)
19. $T = \sin(L_p - \Omega_p) * \sin \iota_p$ where Ω_p is the planet's ecliptic longitude of the ascending node and ι_p is the planet's orbital inclination.
(Ans: $\Omega_p = 113.4888341°$, $\iota_p = 2.4893741°$, $T = 0.038493$)
20. $\Lambda_p = \sin^{-1} T$ where Λ_p is the planet's heliocentric latitude.
(Ans: $\Lambda_p = 2.206024°$)
21. $y = \sin(L_p - \Omega_p) * \cos \iota_p$.
(Ans: $y = 0.885400$)
22. $x = \cos(L_p - \Omega_p)$.
(Ans: $x = -0.463234$)
23. Compute
$$t = \tan^{-1}\left(\frac{y}{x}\right).$$
(Ans: $t = -62.381778°$)
24. Add a quadrant adjustment factor to t to remove the ambiguity of the inverse tangent function.
(Ans: *Adjustment* $= 180°$, $t = 117.618222°$)
25. $L' = \Omega_p + t$.
(Ans: $L' = 231.107056°$)
26. $R' = R_p * \cos \Lambda_p$.
(Ans: $R' = 9.931293°$)
27. $z = L_e - L'$.
(Ans: $z = -130.651512°$)

Steps 28-32 are for an outer planet only. Refer to the previous algorithm for an inner planet.

28. $y = R_e * \sin(-z)$.
(Ans: $y = 0.746011$)
29. $x = R' - R_e * \cos(z)$.
(Ans: $x = 10.571866$)
30. Compute
$$B = \tan^{-1}\left(\frac{y}{x}\right).$$
(Ans: $B = 4.036424°$)
31. Adjust B to account for the ambiguity of the inverse tangent function.
(Ans: No adjustment needed.)
32. $\lambda_p = L' + B$.
(Ans: $\lambda_p = 235.143480°$)
33. If necessary, adjust λ_p to the range $[0°, 360°]$.
(Ans: no adjustment)
34. $y = R' * \tan \Lambda_p * \sin(\lambda_p - L')$.
(Ans: $y = 0.026929$)
35. $x = R_e * \sin(-z)$.
(Ans: $x = 0.746011$)
36. Compute
$$\beta_p = \tan^{-1}\left(\frac{y}{x}\right).$$
(Ans: $\beta_p = 2.067340°$)

37. Convert the ecliptic coordinates λ_p and β_p to their corresponding equatorial coordinates.
(Ans: $\alpha_p = 15^h33^m$, $\delta = -17°3'$)

38. Finally, convert the equatorial coordinates to the appropriate horizon coordinates.
(Ans: $h_p = -65°59'$, $A_p = 33°0'$)

After calculating the number of days since 1/0.0/1980, the algorithms for locating the Sun and Moon added

$$\frac{GMT}{24}$$

to account for the fractional part of the day. This could be done here in step 4 if desired, but the extra effort will do little to improve the accuracy. Compared to the Sun and Moon, the equatorial coordinates of the planets, excepting Mercury and Venus, change very little within a single day.

The astute reader has probably wondered why the gravitational attraction of various planets is not considered in the steps above. To be completely correct, various orbital perturbations must be considered to calculate a planet's location with a high degree of accuracy. This is especially true for the giant planets Jupiter and Saturn which can account for a discrepancy of as much as 1 degree of arc. But since the purpose of this book is not arcsecond accuracy, such considerations are ignored. If such accuracy is desired, orbital corrections would need to be added for each of the planets in much the same manner that orbital corrections were considered during the calculation of the Moon's position.

9.11 Planet Rise and Set Times

The examples worked out in the previous section indicate why it is useful to know when a planet will rise and set. After all that work, neither Venus nor Saturn would be visible to our observer! Calculating the times for Sunrise and Sunset as well as the times for Moonrise and Moonset required an interpolation scheme since the equatorial coordinates of the Sun and Moon change so rapidly. Fortunately, this is not the case with the planets. Calculating the position of a planet once during the day is sufficient to determine its rise and set times.

To illustrate, let us again refer to the observer from the previous section. When will Saturn appear above the observer's horizon? (Note: The date required here is actually the GMT date. Thus, we shall use 12/30/1984 as the date and hence the ecliptic location calculated here will differ slightly from the previous section. This is because the previous section calculated the location for LMT date of 12/30/1984 but a GMT date of 12/31/1984.) The steps are as follows.

1. Calculate the ecliptic location of the desired planet.
(Ans: from the previous section,
$\lambda_p = 235.047797°$, $\beta_p = 2.065590°$)
2. Convert the ecliptic coordinates to equatorial coordinates.
(Ans: $\alpha_p = 15.547930^h$, $\delta_p = -17.024690°$)
3. Using the equatorial coordinates from the previous step, use the algorithm from Chapter 6 to compute the LST times for rising and setting.
(Ans: $LST_r = 10.470675^h$, $LST_s = 20.625185^h$)
4. Finally, convert the LST times to LMT times.
(Ans: $LMT_r = 4^h4^m$, $LMTs = 14^h12^m$)

9.12 Planetary Distance and Angular Size

Calculating the distance to a planet falls out rather easily from the calculation of the planet's location. The equation required is

$$Dist = \sqrt{R_e^2 + R_p^2 - 2 * R_e * R_p * \cos(L_p - L_e)}. \quad (9.12.1)$$

To illustrate, consider calculating the distance to Saturn on December 30, 1984 (GMT date). The resulting distance will be the distance from the Earth to Saturn, not the distance from the Sun to Saturn.

1. Calculate the ecliptic location of the desired planet.
(Ans: $\lambda_p = 235.047797°$, $\beta_p = 2.065590°$)
2. From step 1, obtain R_p, R_e, L_p, and L_e.
(Ans: $R_p = 9.938452$ AU, $R_e = 0.983307$ AU, $L_p = 231.053952°$, $L_e = 99.437015°$)
3. Compute
$$T = R_e^2 + R_p^2 - 2 * R_e * R_p * \cos(L_p - L_e).$$
(Ans: $T = 112.720543$ AU squared)
4. Compute the distance, $Dist = \sqrt{T}$.
(Ans: $Dist = 10.617$ AU)
5. If desired, convert the distance to miles.
(Ans: $Dist = 9.863E08$ miles)

Calculating the angular diameter requires the distance just calculated. The formula required is

$$\theta_p = \frac{\theta_0}{Dist} \quad (9.12.2)$$

where θ_0 is the angular diameter when the desired planet is 1 AU from the Earth and $Dist$ is the distance from the Earth measured in AU. For the planet Saturn on the given date we have

$$\theta_p = \frac{165.6}{10.616990} = 16''$$

or, in DMS format, $\theta_p = 0°0'16''$.

Before the discovery of Neptune and Pluto, astronomers had a rather neat formula for estimating the distance from the Sun to a planet. This rule, called Bode's law, was described by the German astronomer Johann E. Bode in 1772. Even though Bode is given credit for the rule, it was actually worked out by another German astronomer, Johann Daniel Titius, in 1766. In any event, Bode's law is a geometric progression which begins with the number 4. The next number in the progression is given by the equation

$$P_i = 4 + 3 * 2^{i-1}, \quad \text{for} \quad i = 1, 2, 3, \ldots. \qquad (9.12.3)$$

This generates the sequence

4, 7, 10, 16, 28, 52, 100, 196, 388, 772, etc.

Now divide each number in the series by 10 and compare the result with the mean distance to a planet expressed in A.U. units. The results are in Table 9.3 with the column labeled Bode's law being divided by 10. Notice how well Bode's law agrees with the planets from Mercury through Uranus but does not agree very well for Neptune and not at all for Pluto.

At first astronomers paid little attention to Bode's law. Then in 1781 William Herschel discovered Uranus in about the place Bode's law predicted it would be. Other astronomers quickly assumed that there might be something to Bode's law after all and began looking for a missing planet between Mars and Jupiter as predicted by Bode's law. They found an asteroid, Ceres, in 1801 at a distance of about 2.77 A.U. However, with the discovery of Neptune in 1846, the Bode's law theory was abandoned. Pluto, discovered in 1930, dealt a final death blow to Bode's law.

9.13 Perihelion and Aphelion

As a planet moves in its orbit, the distance between the planet and Earth varies. The point in the orbit when the planet is closest to the Earth is called perigee. The point most distant from the Earth is called apogee. Instead of perigee and apogee, what is usually of interest is the times of perihelion and aphelion. These are the points in a planet's orbit when it is closest to or farthest away from the Sun respectively.

Table 9.3
Bode's Law

Planet	Distance (A.U.)	Bode's Law
Mercury	0.3871	0.4
Venus	0.7233	0.7
Earth	1.0000	1.0
Mars	1.5237	1.6
Asteroids		2.8
Jupiter	5.2037	5.2
Saturn	9.5809	10.0
Uranus	19.1168	19.6
Neptune	30.1157	38.8
Pluto	39.3133	77.2

For Mercury, Venus, Earth, Mars, Jupiter, and Saturn the times of perihelion and aphelion are easily calculated. First, a variable K is calculated which is either an integer for perihelion calculations or an integer plus 0.5 for aphelion calculations. Using this value of K, which depends upon the planet, the Julian date can be calculated which corresponds to the desired aphelion or perihelion times. The method used here is not extremely accurate but is sufficient for calculating an approximate time. The error may be as much as a month or so for Saturn, but generally much less for the other planets.

The formula for calculating the variable K and the corresponding formula for the Julian date depend upon the planet under consideration. The formulas given here are from *Astronomical Formulae for Calculators* and are rough approximations that may be incorrect by several weeks.

Mercury

$$\begin{aligned} K &= 4.15201 * (Year - 1900) \\ JD &= 2,414,995.007 + 87.96934997 * K \end{aligned}$$

Venus

$$\begin{aligned} K &= 1.62549 * (Year - 1900) \\ JD &= 2,415,112.001 + 224.7008454 * K - \\ &\quad 0.0000000304 * K^2 \end{aligned}$$

Earth

$$\begin{aligned} K &= 0.99997 * (Year - 1900) \\ JD &= 2,415,021.546 + 365.2596413 * K + \\ &\quad 0.0000000152 * K^2 \end{aligned}$$

Mars

$$K = 0.53166 * (Year - 1900)$$
$$JD = 2,415,097.251 + 686.9958091 * K - 0.0000001221 * K^2$$

Jupiter

$$K = 0.08430 * (Year - 1900)$$
$$JD = 2,416,640.884 + 4,332.894375 * K + 0.0001222 * K^2$$

Saturn

$$K = 0.03393 * (Year - 1900)$$
$$JD = 2,409,773.47 + 10,764.18010 * K + 0.0013033 * K^2.$$

The variable $Year$ in these equations must also include the month and day as a fractional part of the year. The value obtained for K is only an approximation. The result must then be rounded to the nearest integer for computing perihelion or to the nearest number ending in 0.5 for aphelion. Given the Julian date, it is easy to compute the calendar date and corresponding GMT. Note that the Julian date which results will be the one closest to the date entered. Hence, the times of aphelion and perihelion may well be significantly later or earlier than the date entered.

Using values of K which are not integers or are not numbers ending in 0.5 will produce meaningless results. Furthermore, greater error is incurred for years that are significantly distant from 1900.

As an example, let us compute the perihelion and aphelion times for Mars closest to October 30, 1938. The steps are:

1. Convert Day to number of days into the year. (Ans: $Days = 303$)
2. Compute
$$y = Year + \frac{Days}{365.25}.$$
(Ans: $y = 1,938.8300$)
3. Compute $T = K_p*(Y-1900)$ where K_p is obtained from the appropriate equation above for calculating the value of K. (Ans: $K_p = 0.531660$ for Mars, $T = 20.644140$)
4. Let K_{per} be the integer value closest to T and let K_{aph} be the fraction ending in 0.5 which is closest to T. (Ans: $K_{per} = 21$, $K_{aph} = 20.5$)
5. Compute the Julian dates for the given planet using K_{per} and K_{aph}. (Ans: $JD_{aph} = 2,429,180.6650$, $JD_{per} = 2,429,524.1629$)
6. Convert the Julian dates from the previous step to calendar dates. (Ans: $JD_{aph} = 10/9.17/1938$, $JD_{per} = 9/17.66/1939$)
7. Convert the days from the previous step to Day and GMT. (Ans: Day_{aph} = day 9, 3.960846^h GMT, Day_{per} = day 17, 15.910492^h GMT)
8. Convert the time to HMS format for the resulting times of perihelion and aphelion. (Ans: *Aphelion*: 10/9/1938 at 3^h58^m GMT, *Perihelion*: 9/17/1939 at 15^h55^m GMT)

9.14 Planet Phases

The planets go through phases just like the Moon does. But instead of computing phases and age, this section will only deal with the percent illumination factor. The percent illumination is

$$K = \frac{(R_p + Dist)^2 - R_e^2}{4 * R_p * Dist} \tag{9.14.1}$$

where $Dist$ is the distance from the Earth to the planet. To illustrate the steps, consider Venus on December 30, 1984.

1. Compute the ecliptic coordinates of the desired planet. (Ans: $\lambda_p = 325.166988°$, $\beta_p = -1.632681°$)
2. Obtain R_p, R_e, L_p, and L_e from the previous step. (Ans: $R_p = 0.723233$ AU, $R_e = 0.983307$ AU, $L_p = 42.050858°$, $L_e = 99.437015°$)
3. Compute the distance from the Earth to the planet using the technique shown earlier. (Ans: $Dist = 0.850508$ AU)
4. Compute $T = (R_p + Dist)^2 - R_e^2$. (Ans: $T = 1.509770$)
5. Calculate
$$K = \frac{T}{4 * R_p * Dist}.$$
(Ans: $K = 0.613612$)
6. $Percent\ Illumination = 100 * K$. (Ans: $Percent\ Illumination = 61\%$)

9.15 Miscellaneous Calculations

With the completion of the space shuttle, space travel has the promise of becoming a little more commonplace. And with new projects like the space telescope, the space station, and the proposed Galileo probe on the horizon, space exploration may be even more dramatic in the very near future. But most of us will not have the opportunity to actively participate in these adventures. However, as a poor man's substitute, we can at least look at some of the interesting factors that have to be considered by those who will. We will start by looking at how much an astronaut would weigh on one of the planets. Then we will calculate how long it takes light to reach us from a planet to see what propagation delays are involved in radio transmissions. Just for fun, we will also calculate the length of a year on the planets. And finally, one of the most important calculations of all is to determine how fast a spacecraft must travel to be able to leave a planet and head back home to Earth.

9.15.1 Weight on a Planet

Newton's *Law of Universal Gravitation* provides a way to determine how much an object will weigh on another planet. Newton's law says that all objects in the universe are attracted to each other. The force of attraction between them is directly proportional to the product of their masses and inversely proportional to the square of the distance between them. Mathematically, this can be expressed as

$$F = \frac{g * m_1 * m_2}{R^2} \tag{9.15.1}$$

where g is a gravitational constant, m_1 and m_2 are the masses of the two objects, R is the distance between the two objects, and F is the resulting force. When one of the masses is the Earth, say m_2, the resulting force F is called the weight of the object whose mass is m_1.

Let m be the mass of an astronaut with m_e representing the Earth's mass and let m_p represent the mass of a planet. Furthermore, let R_e and R_p represent the radius of the Earth and the planet respectively. Let W_e and W_p represent the astronaut's weight on the Earth and on the planet. Applying Newton's law give us

$$W_e = \frac{g * m * m_e}{R_e^2} \tag{9.15.2}$$

and

$$W_p = \frac{g * m * m_p}{R_p^2}. \tag{9.15.3}$$

Dividing W_p by W_e will produce the ratio of the astronaut's weight on the planet to his weight on the Earth. That is,

$$\begin{aligned} \frac{W_p}{W_e} &= \frac{g * m * m_p}{R_p^2} \div \frac{g * m * m_e}{R_e^2} \qquad (9.15.4) \\ &= \left(\frac{g * m * m_p}{R_p^2}\right)\left(\frac{R_e^2}{g * m * m_e}\right) \\ &= \frac{m_p * R_e^2}{R_p^2 * m_e} \end{aligned}$$

Notice that the ratio only depends on the planet's size and mass and the Earth's size and mass. This should not be too surprising since it says that the percentage by which one astronaut's weight is affected is the same as another regardless how much they weigh.

As an example, how much did Neil Armstrong weigh on the Moon? Assume that on Earth Mr. Armstrong weighs 175 pounds. From Table 9.2, we find that $m_e = 1$, $m_p = 0.01228$, $R_e = 6,371$ km, and $R_p = 1,738$ km. (Actually, Table 9.2 gives masses relative to Earth. It makes no difference in working the problem.) Using these values to obtain the weight ratio, we obtain

$$\frac{W_p}{W_e} = \frac{(0.01228)(6,371)^2}{(1)(1,738)^2} = 0.165.$$

This ratio says that weight on the Moon is 16.5% what it is on the Earth. Since we assumed an Earth weight of 175 pounds, the weight on the Moon is

$$W_p = (0.165)(175) = 29 \text{ pounds}.$$

Once the weight factor ratio is known for a planet, multiply the ratio by the Earth weight to obtain how much the object weighs on the planet.

9.15.2 Radio Transmission Delays

During the Voyager missions, there was a noticeable delay from the time a command was sent from the Earth until the space probe received the command. This is because radio waves travel at a finite speed. When an object is far away, the delay can be very noticeable. Let us find out how much the delay is.

Suppose an object is traveling at s miles per second for t seconds. The distance traveled, in miles, is given by

$$Dist = st. \tag{9.15.5}$$

What we seek is how long it takes to travel some distance, so solving for t we have

$$t = \frac{Dist}{s}. \tag{9.15.6}$$

To make future calculations easier, let us find out how long it takes for light to travel 1 A.U. The speed of light is about 186,400 miles per second while, from Chapter 2, 1 A.U. is equal to about 9.29E7 miles. Substituting these values into our equation, we have

$$t = \frac{9.29\text{E miles}}{186,400 \text{ miles/sec}} = 498 \text{ seconds}.$$

Converting seconds to hours, we find that light travels 1 A.U. in about 0.1384 hours. Finding out how long it takes light to reach the Earth from another planet is simple to calculate. Multiply the distance from Earth to the planet (measured in astronomical units) by 0.1384 as indicated in the formula

$$t = 0.1384 * Dist. \tag{9.15.7}$$

Suppose a space probe reached Saturn on December 30, 1984. We have already calculated the distance from Saturn to the Earth on that date to be 10.62 A.U. Multiplying this by 0.1386 will indicate how long it takes for light to reach Earth from Saturn. The result is 1.469^{h} or, converting to HMS format, $1^{\text{h}}28^{\text{m}}$. Thus, any message sent from the Earth to the probe would take $1^{\text{h}}28^{\text{m}}$ to get there. A reply from the probe to Earth would also take $1^{\text{h}}28^{\text{m}}$ so that if a command is sent which requires a reply, there would be a total delay of $2^{\text{h}}56^{\text{m}}$.

9.15.3 Length of a Planetary Year

As described in Chapter 4, a year is the time it takes for Earth to revolve around the Sun. In a similar manner, a year for a planet can be defined to be the time it takes for a planet to go around the Sun. Table 9.2 gives this time for each of the planets in terms of tropical years. Since a tropical year is 365.2422 Earth days, the length of a planet's year can be calculated by multiplying its period by 365.2422. Mathematically, we have

$$Year_p = 365.2422 * T_p. \tag{9.15.8}$$

For example, Saturn's period is 29.45771 tropical years. Multiplying this by 365.2422 gives Saturn's year as 10,759.198807 Earth days. This is equivalent to 29 years, 167 days, $4^{\text{h}}12^{\text{m}}$.

9.15.4 Escape Velocity

Our final calculation is to determine the escape velocity for a planet. The escape velocity is the speed necessary for an object to be able to escape the gravitational field of a planet. From Newton's *Law of Universal Gravitation*, it is clear that the force acting upon an object is greater for a more massive planet than for a smaller one. Hence, the velocity required to leave the more massive planet is greater than for a smaller one. Without motivation, the equation for the escape velocity is

$$E_v = \frac{\sqrt{2 * g * m_p}}{\sqrt{R_p}}. \tag{9.15.9}$$

where E_v is the escape velocity in meters per second, m_p is the planet's mass in kilograms, and R_p is the planet's radius in meters. The gravitational constant is

$$g = 6.67\text{E-11 Newton-m}^2/\text{kg}^2.$$

(The meaning of the gravitational constant units need not concern us.)

As an example, let us calculate the Moon's escape velocity. From Table 9.2, the Moon's radius is 1738 km so that R_p is 1.738E6 meters. The Moon's mass relative to the Earth is 0.01228. Now the Earth's mass is 5.98E24 kilograms so that

$$m_p = (0.01228)(5.98\text{E}24) = 7.34344\text{E}22 \text{ kilograms}.$$

Substituting these values, we have

$$\begin{aligned} E_v &= \frac{\sqrt{(2)(6.67\text{E}-11)(7.34344\text{E}22)}}{\sqrt{1.738\text{E}6}} \\ &= 2374 \text{ meters/sec} \\ &= 2.374 \text{ km/sec} \end{aligned}$$

This is equivalent to 1.475 miles per second or 5310 miles per hour. By comparison, the Earth's escape velocity is 11.19 km/sec or 25031 miles/hour.

9.16 About the Program

The last program, PLANETS, is the culmination of the entire book. With this program, and portions of MOON and SUN, the location of the Sun, Moon, planets, and stars can be calculated. The reader is encouraged to combine the subroutines as desired to produce custom programs. One such program might be to calculate the location for each planet, the Sun, the Moon, and the rising and setting times for each of these. You might want to include the subroutine from program STARS which plots star locations to include the Sun, Moon, and planets to produce your own star charts and to create your own astronomical calendar of celestial events.

There is one more chapter past this one although this chapter represents the climax of the book. The road to this chapter and to this last program may have been long, but certainly it has been well worth the effort. You, the reader, have at your fingertips the ability to make predictions and calculations about our solar system that the ancients only dreamed about. With a little effort, you can combine pieces from various programs to customize the software to meet your own needs. Experiment and enjoy!

9.17 Sample Problems

1. The date is January 22, 1985 at 22^{h} LMT. Assume an observer is not on Daylight Saving Time but is within the Eastern Standard Time Zone. If his location is 78.3°W longitude, 37.8°N latitude, calculate the equatorial and horizon coordinates for

Table 9.4 Weight, Time, Year, and Escape Velocity				
	Weight (relative to Earth)	Time (Earth hrs)	Year (Earth days)	Escape Velocity (miles/sec)
Sun	27.96			383.69
Mercury	0.38	0^h8	0y 87d 23^h15^m	2.64
Venus	0.91	0^h7^m	0y 224d 16^h49^m	6.44
Earth	1.00		1y 0d 0^h21^m	6.95
Mars	0.38	0^h14^m	1y 321d 17^h43^m	3.12
Jupiter	2.54	0^h51^m	11y 314d 22^h14^m	37.01
Saturn	0.84	1^h28^m	29y 167d 4^h12^m	20.75
Uranus	0.90	2^h46^m	84y 4d 13^h19^m	13.20
Neptune	1.19	4^h20^m	164y 290d 13^h54^m	14.76
Pluto	0.08	4^h11^m	247y 0d 0^h0^m	0.83
Moon	0.17			1.47

Mercury and Jupiter.
(Ans: For Mercury, $\alpha = 19^h10^m$, $\delta = -23°20'$, $h = -68°39'$, $A = 307°24'$.
For Jupiter, $\alpha = 19^h55^m$, $\delta = -21°7'$, $h = -59°33'$, $A = 294°48'$)

2. For the previous problem, calculate the rising and setting times for Mercury and Jupiter.
(Ans: For Mercury, $LMT_r = 6^h28^m$, $LMT_s = 15^h50^m$.
For Jupiter, $LMT_r = 7^h10^m$, $LMT_s = 16^h48^m$)

3. For problem number 1, calculate the distance, angular diameter, and percent illumination for Mercury and Jupiter.
(Ans: For Mercury, $Dist = 1.3$ AU, $\theta = 5.0''$, and % Illum = 89.
For Jupiter, $Dist = 6.1$ AU, $\theta = 32.0''$, and % Illum = 100)

4. Calculate the times of perihelion and aphelion closest to January 22, 1985 for the planets Mercury, Venus, Mars, Jupiter, and Saturn.
(Ans: see Table 9.5)

5. Assuming the date is December 30, 1984, calculate the weight factor, time for light to reach the Earth, the length of a year in Earth days, and the escape velocity for each of the planets, including the Earth, as well as the Sun and Moon.
(Ans: see Table 9.4)

Table 9.5 Planet Perihelion and Aphelion		
	Perihelion	Aphelion
Mercury	12/13/84 16^h30^m	1/26/85 16^h8^m
Venus	2/24/85 5^h13^m	6/16/85 13^h37^m
Mars	11/7/84 13^h30^m	10/17/85 1^h26^m
Jupiter	6/24/87 15^h37^m	5/30/93 2^h22^m
Saturn	1/18/74 12^h32^m	4/25/59 10^h17^m

Chapter 10

Useful Astronomical Aids

The major emphasis of this book has been to demonstrate the calculations needed to predict the location of various celestial objects. Such calculations are long and tedious, but with the aid of a microcomputer, the tediousness is eliminated. Even so, there are times when it may be more convenient to refer to some other source for information than it is to calculate it yourself. This is especially true if great precision is required or if you do not have the formulas you need to calculate the information you want. Thus, this chapter describes some common observational aids that can prove to be very useful and time saving. Hopefully you will find that this book and its programs are your most valuable and time saving resource!

There are a great number of observational aids available in the form of charts or tables. Some are printed on a monthly or annual basis while others are referenced to some standard epoch. Most provide information on the location of an object, generally in equatorial coordinates, but may also provide other interesting information such as magnitude and distance. Unless great precision is needed, you may find that data provided in monthly periodicals such as *Sky & Telescope* are exactly what you need. This is especially true for general "sight seeing" when all you really want is a general idea of where to point the telescope.

When using the various astronomical aids, it should be obvious that they are not likely to be adjusted for your exact location or time zone. Some aids assume a location in the middle of the United States, but the more precise ones assume 0^h GMT at 0° latitude, 0° longitude. In any event, once the equatorial coordinates of an object are determined, the results of Chapters 4 and 5 can be used to calculate the horizon coordinates for your location while Chapter 6 can be used to determine rising and setting times.

To avoid confusion, some definitions are required. The terms star chart and star map will be used interchangeably to refer to a single sheet of paper which contains a pictorial representation of the relative position of celestial objects. The representation may be for the entire night sky as seen from some vantage point, or may only be an expanded view of a small section of the sky. A star atlas will refer to a bound collection of several star charts. The key thing to remember is that star charts, maps, and atlases are pictorial representations of the sky. Note that even though they are called star charts, such pictures may also include galaxies, nebulae, and even planets.

Catalogues, ephemerides, tables, and almanacs are often used interchangeably since the main feature of each is that they give a numeric indication (typically equatorial coordinates) of the location of an object. A catalogue (as in star catalogue) typically gives the location of stars, galaxies, quasars and other objects which lie outside the solar system. A star catalogue lists the location for some specific epoch and precession adjustments must be made, as shown in Chapter 5, to convert to another epoch. On the other hand, an ephemeris (a book containing ephemerides), astronomical table, or almanac typically refers to objects within the solar system. Moreover, these tables typically give the position of an object for several instants during the year, perhaps even a daily position for the object. The main point here is that such aids give the coordinate location of an object as opposed to just a pictorial representation of relative locations.

This chapter will begin with a look at star maps and how to use them. A set of star maps, such as Wil Tirion's *Sky Atlas 2000.0*, is a good addition to your observatory and should be an early purchase since they are so useful for planning a night of observations. After discussing such maps, we will then discuss ephemerides which give the location of various objects. Finally, the chapter concludes with a discussion of astronomical calendars which graphically depict many of the more interesting astronomical events.

10.1 Star Maps

A star map contains a pictorial representation of the night sky. While some are more detailed than others, the basic idea is to represent what the night sky looks like from some particular viewpoint. First, consider the star maps published monthly in some of the popular

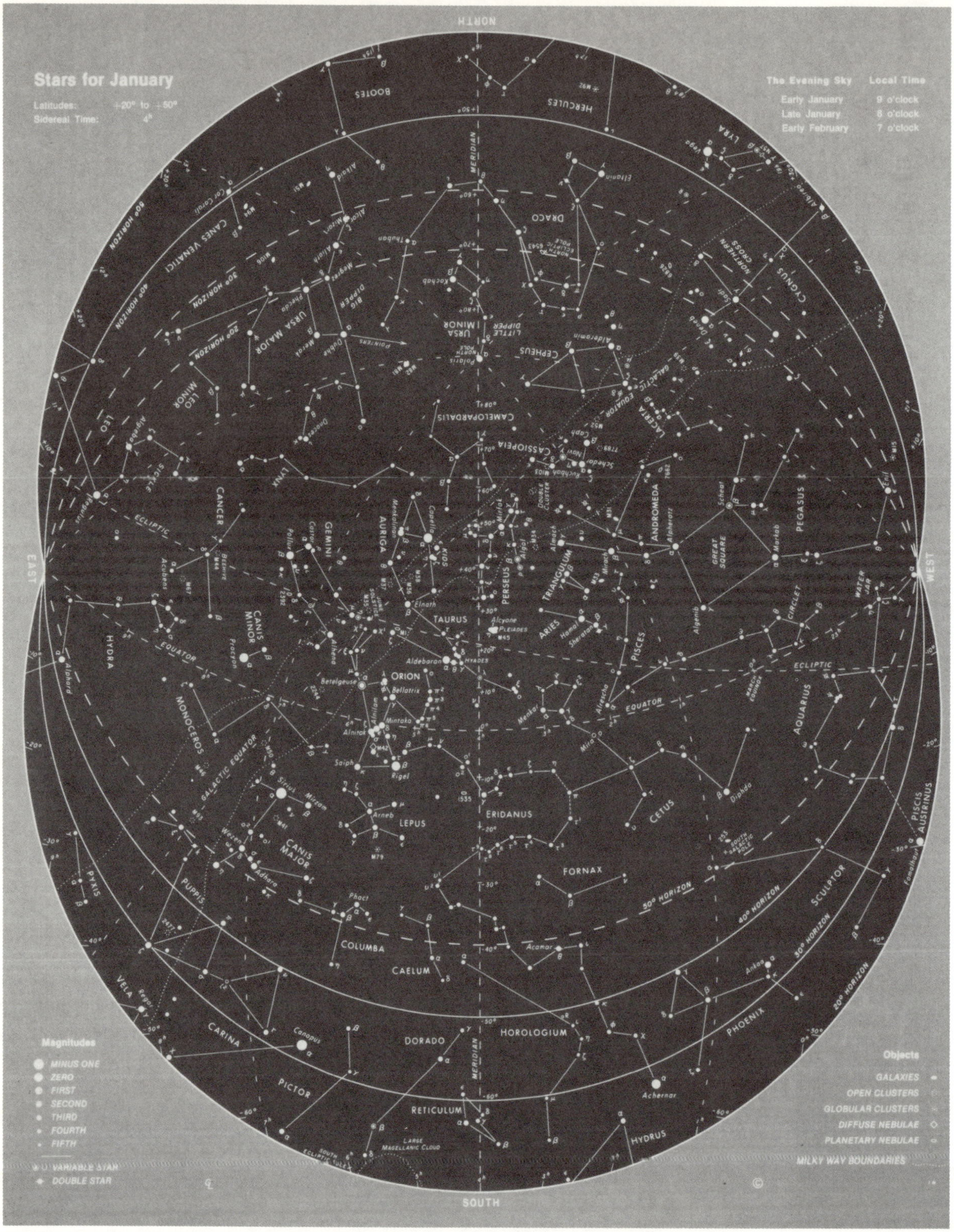

Figure 10.1 *This star chart, from* Sky & Telescope, *contains horizon circles for different latitudes in the Northern Hemisphere as an aid in determining what stars will be visible for a particular location. ©1967, Sky Publishing Corp.*

astronomy periodicals. Since they are published on a monthly basis, the maps may already be adjusted to account for precession, but generally the resolution is so low that it makes no practical difference. To utilize such star maps, use a compass to locate magnetic North and align the map's North direction (typically the top) with magnetic North. Select some easily recognizable constellation, such as the Big Dipper, on the chart as a reference point and then try to find the constellation in the night sky. Lines may be drawn on the chart to connect stars in a constellation to make recognizing them easier.

When looking at a star chart which has the compass directions indicated on them, be careful to note the location of East and West. Such maps are frequently printed with East and West reversed so as to present a view of the sky as if you were lying on your back and looking up into the night sky.

Depending upon your latitude, you may not be able to see all of the stars indicated on the star map. Maps printed on a monthly basis will often have horizon circles printed on them as a guide for determining what stars you will be able to see. For example, if your latitude is greater than 30°, you will not be able to see any of the stars which lie outside the 30° horizon circle printed on the map.

In addition to having horizon circles, star maps from periodicals will often have lines which form a grid to give a rough indication of the equatorial coordinates of an object within the grid. One side of the grid will indicate the declination while the other side will indicate the right ascension. The right ascension is often expressed as the hour angle since most equatorial mount telescopes have hour circles on them.

Star charts such as those from *Sky Atlas 2000.0* are much more detailed than that described so far. You may find initially that the added detail makes the charts somewhat more difficult to use. But after some practice, they will become second nature to you and will provide much more information than the less detailed maps. These star charts will usually give the star's magnitude by making brighter stars larger than fainter ones.

A star atlas will be referenced to some standard epoch. But this may be of no practical importance since coordinates can only be estimated within a few minutes depending upon the grid spacing used to measure the equatorial coordinates. Since compass directions are not noted on such maps, you will need to use the results of Chapter 5 to see where the map section of interest will appear for your location. Pick a sample equatorial coordinate value from the chart, calculate the corresponding horizon coordinate system values, and you will be able to tell where in the sky to look for objects on that section of the star chart. Moreover, you will need the results of Chapter 6 to determine if that section of the sky will even be visible for the time you choose for observation.

Star maps are often useful for locating objects other than stars. For example, suppose that you know that Saturn is going to be in the constellation Libra at some particular time. By using the star map as a guide, you can locate the constellation of Libra in the night sky and probably spot Saturn very easily. Saturn will be the bright "star" which does not appear on your star chart.

10.2 Star Catalogues

If you need to have the precise location of an object outside the solar system, a star catalogue is your best bet. However, to use the star catalogue, you must know the name of what you are looking for. The easiest way to determine an object's designation is to first find it on a star chart. The chart may in fact have two names for the object. For example, if you look on a star chart and find the bright star in the lower right corner of Orion, you will see that it is called Rigel and also called β. In fact, you will find several stars on the chart which are called β ! There are unfortunately several ways to name a star. Let's look at the more common methods that you are likely to encounter.

The key to finding a star is to first determine what constellation it resides in. Roughly speaking, and you should know there are exceptions, the brightest star in the constellation, as measured by its magnitude, is labeled with the first letter in the Greek alphabet, α. The second brightest star is usually labeled with β, the second letter in the Greek alphabet, and so on. Some of the brighter stars in a constellation will even have a proper name such as the star Rigel mentioned above. Thus, since Rigel is labeled β, it is the second brightest star in the Orion constellation. The brightest star in the Orion constellation is Betelgeuse and hence it is labeled α. To locate this star in a star catalogue, you will find it listed with the stars for the constellation Orion and designated as Betelgeuse or α.

Two stars in the same constellation can have the same Greek letter but differentiated by a superscript. For example, if you look just above and to the right of Betelgeuse (Orion constellation) on a detailed star chart, you will find two stars labeled φ^1 and φ^2. Originally, only a few stars were lettered. As astronomers more systematically cataloged the stars many "new" ones were discovered. Some of these were so close to "old" lettered stars and of about the same magnitude that the expedient of superscripts was adopted to avoid confusion.

So far so good, but what happens if a constellation contains more stars than there are letters in the Greek alphabet? Bayer adopted a sequence of Roman letters, both upper and lower case, which was adequate for the relatively few stars plotted on his atlas. The advent of the telescope and man's quest to catalog more stars required another method. Using a French edition of John Flamsteed's *Atlas Coelestis* Joseph Jerome de Lalande numbered the stars. These "Flamsteed" numbers are simply integers assigned to each star by constellation in order of right ascension. For Rigel, the Flamsteed number is 19. Thus, Rigel may be found in a star catalogue in several ways: as Rigel, as Orion- β, and as Orion-19!

The situation becomes even more discouraging for identifying non-stellar or "deep-sky objects." These objects are labeled by a number which depends upon which catalogue you look at! Three common catalogues are the *New General Catalogue* (NGC), the *Index Catalogue, I and II* (IC), and the *Messier Catalogue* (M). On the Wil Tirion star charts, numbers without a letter preceding them are NGC designations, those with an M before them are Messier catalogue designations, while those with an I are IC designations. Proper names may also exist for the object as in the case of the Great Orion Nebula. It appears on the Wil Tirion chart as 1972-M42 indicating that in the NGC catalogue, its designation is 1972 while in the Messier catalogue, it is designated 42. Many star charts label the Great Orion Nebula as NGC-1972 to make it clear which catalogue numbering system is being used.

After looking up a star in a star catalogue to get its equatorial coordinates, you may then use the results from Chapters 5 and 6 to predict where it will appear in the night sky for your location as well as its rising and setting times. But in addition to equatorial coordinates, star catalogues also contain other useful information such as the star's magnitude, distance, and classification (red giant, white dwarf, etc.).

10.3 Ephemerides and Almanacs

An ephemeris or astronomical almanac contains tables which give the location of solar system objects for some time period. The time period may be a monthly, weekly, or even daily basis. In addition, other information may be given such as rising and setting times, times for eclipses, times of the equinoxes and solstices, and times for perihelion and aphelion.

Since the solar system objects move so rapidly across the sky, the equatorial coordinates must be specified with respect to a date and time. The coordinates are normally given for 0^h GMT. Some data, such as rise and set times, are for a specific location which is generally 0° latitude (the equator), 0° longitude (the Greenwich Prime Meridian). In some almanacs, the tables may be calculated for several latitudes in addition to the equator. The time conversions in Chapter 4 can be used to convert the results to your particular LMT.

Using an ephemeris to determine the position of a planet is more complicated than using a star catalogue to locate a star or other deep-sky object. The reason is that due to the relatively rapid motion of the planets, daily corrections must be made to the equatorial coordinates and for very high precision, corrections may need to be made on an hourly basis. Hourly corrections are obviously required for the Sun and the Moon. Two methods frequently used for these calculations are described here. Both methods are based upon the evaluation of a polynomial, but the polynomial to evaluate may differ depending upon the book used as a reference.

The technique used in *Almanac for Computers* is to define a pair of polynomial functions which have time as the independent variable and produce the equatorial coordinates as the dependent variables. Three steps are required. First, the time (GMT) and day are combined by the formula

$$t = \text{Day of the Month} + \frac{GMT}{24}. \qquad (10.3.1)$$

Secondly, this time variable t is normalized to lie in the range $[-1, 1]$ by the formula

$$x = \frac{t - W}{A} - 1. \qquad (10.3.2)$$

The last step is then to use x to evaluate a pair of polynomial functions which will give the desired coordinates. The constants W and A and the polynomial functions to use depend upon the object under consideration and the date.

To illustrate, consider calculating the Moon's declination for June 5, 1986 at 6^h GMT. For the time period from June 1 to June 8, *Almanac for Computers* gives the following values to the constants A and W:

$$A = 4.0, \quad W = 1.0.$$

The polynomial function for the declination contains 8 terms and is given as

$$\begin{aligned} \mathrm{f(x)} \;=\; & 18.0441 + 17.6853x - 5.7634x^2 - \qquad (10.3.3) \\ & 2.0237x^3 - 0.0898x^4 + 0.0924x^5 + \\ & 0.0710x^6 - 0.0040x^7. \end{aligned}$$

Actually, two sets of polynomials for the equatorial coordinates are given in the almanac. The one shown here is sufficiently accurate for navigational purposes. Another polynomial is given for astronomical uses, but the polynomial requires an evaluation of 36 terms while a similar one for the planets requires evaluation of 38 terms! The process required for either polynomial is the same, but we show the shorter one here since it requires less effort and is precise enough for our purposes.

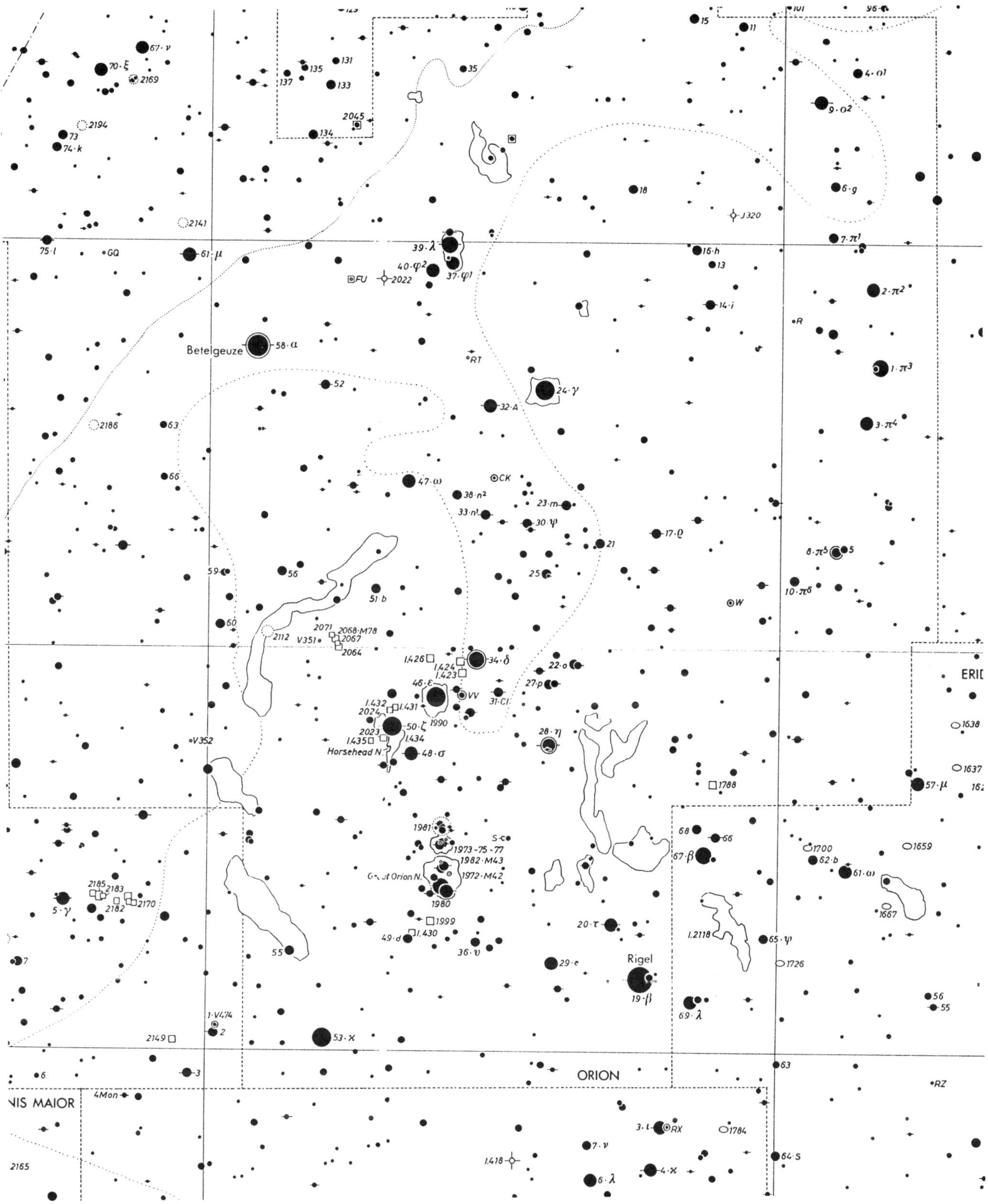

Figure 10.2. *A portion of the constellation Orion as depicted in Wil Tirion's Sky Atlas 2000.0. Both stars and "Deep-Sky" objects can be labeled in a variety of ways. Note that the bright star in the lower right hand corner of Orion has three "labels": the name Rigel, a Bayer letter β and a Flamsteed number of 19. Similarly the non-stellar object forming part of Orion's Sword can be identified as New General Catalogue (NGC) 1982 or as Messier (M) 43. ©1981, Sky Publishing Corp.*

PLANETARY DATA FOR JANUARY, 1986

Object	Date	R.A. h m	Dec. ° ′	Elong. °	Mag.	Diam. ′ ″	Illum. %	Dist. a.u.
Sun	Jan 1	18 44.7	−23 03	—	−26.8	32 35.0	—	0.983
	16	19 50.1	−21 02	—	−26.8	32 34.2	—	0.984
	31	20 53.1	−17 32	—	−26.8	32 31.3	—	0.985
Mercury	Jan 1	17 29.1	−22 58	17 Mo	−0.4	5.3	88	1.273
	16	19 07.9	−23 54	10 Mo	−0.6	4.8	96	1.401
	31	20 52.4	−19 44	2 Mo	−1.3	4.8	100	1.408
Venus	Jan 1	18 25.4	−23 39	4 Mo	−3.9	9.8	100	1.703
	16	19 47.1	−22 03	1 Mo	−3.9	9.7	100	1.711
	31	21 05.5	−17 58	3 Ev	−3.9	9.8	100	1.710
Mars	Jan 1	14 33.9	−14 01	60 Mo	+1.4	4.9	93	1.897
	31	15 46.5	−19 04	72 Mo	+1.1	5.8	91	1.614
Jupiter	Jan 1	21 23.8	−16 07	38 Ev	−2.0	34.1	—	5.774
	31	21 50.5	−13 56	14 Ev	−2.0	32.9	—	5.977
Saturn	Jan 1	16 14.3	−19 23	35 Mo	+0.5	15.4	—	10.760
	31	16 25.7	−19 48	63 Mo	+0.5	15.9	—	10.385
Uranus	Jan 16	17 17.9	−23 09	35 Mo	+5.7	3.5	—	19.916
Neptune	Jan 16	18 17.9	−22 19	21 Mo	+8.0	2.2	—	31.162
Pluto	Jan 16	14 40.6	+1 42	79 Mo	+13.7	0.1	—	29.911

Table 10.1: *This table for the month of January, 1986, shows how* Sky & Telescope *magazine provides positional and other planetary information each month. ©1985 Sky Publishing Corp.*

Substituting in the known values for W, A, the day of the month, the time, and then performing the computations indicated, we have

$$t = 5 + \frac{6}{24} = 5.25,$$

$$x = \frac{5.25 - 1}{4.0} - 1 = 0.0625.$$

Now substituting x in the given polynomial and evaluating it, we have

$$\delta = f(0.0625) = 19.1264^\circ = 19^\circ 7' 35''.$$

A similar polynomial would be evaluated to arrive at the hour angle, H, or the right ascension, α.

A simpler, but less precise, method can be used with the ephemerides printed in many of the popular astronomy periodicals such as *Sky & Telescope*. The main idea is that if positions are given for an object at two different times, an interpolation can be done to determine where the object is at times within the given time interval. That is, suppose the position is A at day D_a and position B at day D_b. Let d be the date of interest, noting that d must lie between D_a and D_b. Armed with this information, two steps are required. First, calculate the number of days from D_a to d with the formula

$$t = d - D_a. \tag{10.3.4}$$

Secondly, evaluate the polynomial equation

$$f(t) = \frac{t(B - A)}{D_b - D_a} + A. \tag{10.3.5}$$

As an example, calculate the right ascension of the planet Mercury on January 12, 1986. *Sky & Telescope* shows that the right ascension for January 1 is $17^h 29.1^m$ and is $19^h 7.9^m$ for January 16. Thus, we have the following values:

$D_a = 1, D_b = 16, d = 12$
$A = 18^h 44.7^m = 17.485^h$
$B = 19^h 50.1^m = 19.1317^h$.

Using these values, we calculate t as

$$t = 12 - 1 = 11$$

and $\alpha = f(t) = f(11)$ as

$$f(11) \quad = \quad \frac{11(19.1317 - 17.485)}{16 - 1} + 17.485$$

$$= 18.6926$$
$$= 18^h41^m33^s.$$

In a similar fashion, the declination could be computed for Mercury.

This technique is obviously faster than the previous method, but a penalty is paid in terms of accuracy. Still, for manually pointing a telescope, the accuracy is sufficient. If one is willing to sacrifice accuracy, the process can be simplified even further. In the example cited for Mercury, one could use B ($19^h7.9^m$) as the right ascension since that is the value for the date closest to January 12. The resulting difference is 26^m21^s which may be close enough, depending upon what you need the right ascensionfor. Still, you may find it worth the extra few minutes to carry out the interpolation.

10.4 Astronomical Calendars

If one is willing to sacrifice accuracy and is only concerned with time and not equatorial coordinates, one of the most useful astronomical aids is an astronomical calendar. One such calendar, the *Sky-Gazer's Almanac*, is published annually in the *Sky & Telescope* magazine. Since this calendar is readily available and is representative of such devices, it will be the basis of the description below on how to use an astronomical calendar.

An astronomical calendar is a very useful and versatile graphical time chart which contains a pictorial representation of the major astronomical events which will occur throughout the year. At a glance it is possible to estimate events such as the following, among may others:

- Times of Sunrise and Sunset
- Times of Moonrise and Moonset
- Phase of the Moon
- Times when planets rise and set
- When planets are in conjunction or opposition
- Transit times for major stars such as Vega
- When meteor showers are most likely to occur

In addition to estimating times for certain astronomical events, the astronomical calendar may also provide a quick way for calculating Julian dates, sidereal time, and the Equation of Time.

The astronomical calendar lists calendar dates along the vertical axis of the chart while the time is marked off along the top and bottom of the chart. While dates are numbered a week apart, the vertical lines of the dotted gridwork each contain dots for each intermediate day. Similarly, the horizontal lines contain dots for each 5 minutes of time. The vertical lines are spaced half an hour apart; the horizontal ones are one week apart. But the intermediate dots on both sets of lines enable the event curve to be read with relatively high accuracy, more so than often imagined, for such a graphical calendar.

Events which occur on the calendar are for a specific longitude and latitude (90°W, 40°N for the *Sky Gazer's Almanac*) so that corrections to the time are generally required. Specific instructions will be included with the calendar, but the basic idea is to account for Daylight Saving Time and then account for time zone differences.

Reading an astronomical calendar is very simple. Locate the date along the vertical axis of the chart and read across the chart until you reach an "event line." An event line is an indication that some astronomical event occurs such as the rising of a planet. Having reached an event line for a given date, the time for the event is found by looking directly above (or below as convenient) the event line to the time marked along the horizontal axis. This will give the LMT that the event is to occur but as explained in the previous paragraph, you may need to make corrections depending upon your specific location.

Besides being simple to use, the astronomical calendar graphically depicts the comings and goings of celestial bodies throughout the year. This will give you a better feel for the rhythmic motions of the Sun, Moon, and planets. You may wish to use the programs in this book to create an astronomical calendar yourself for your specific location to give you an even better insight into the "music of the spheres" as you compare your calculations against actual night observations.

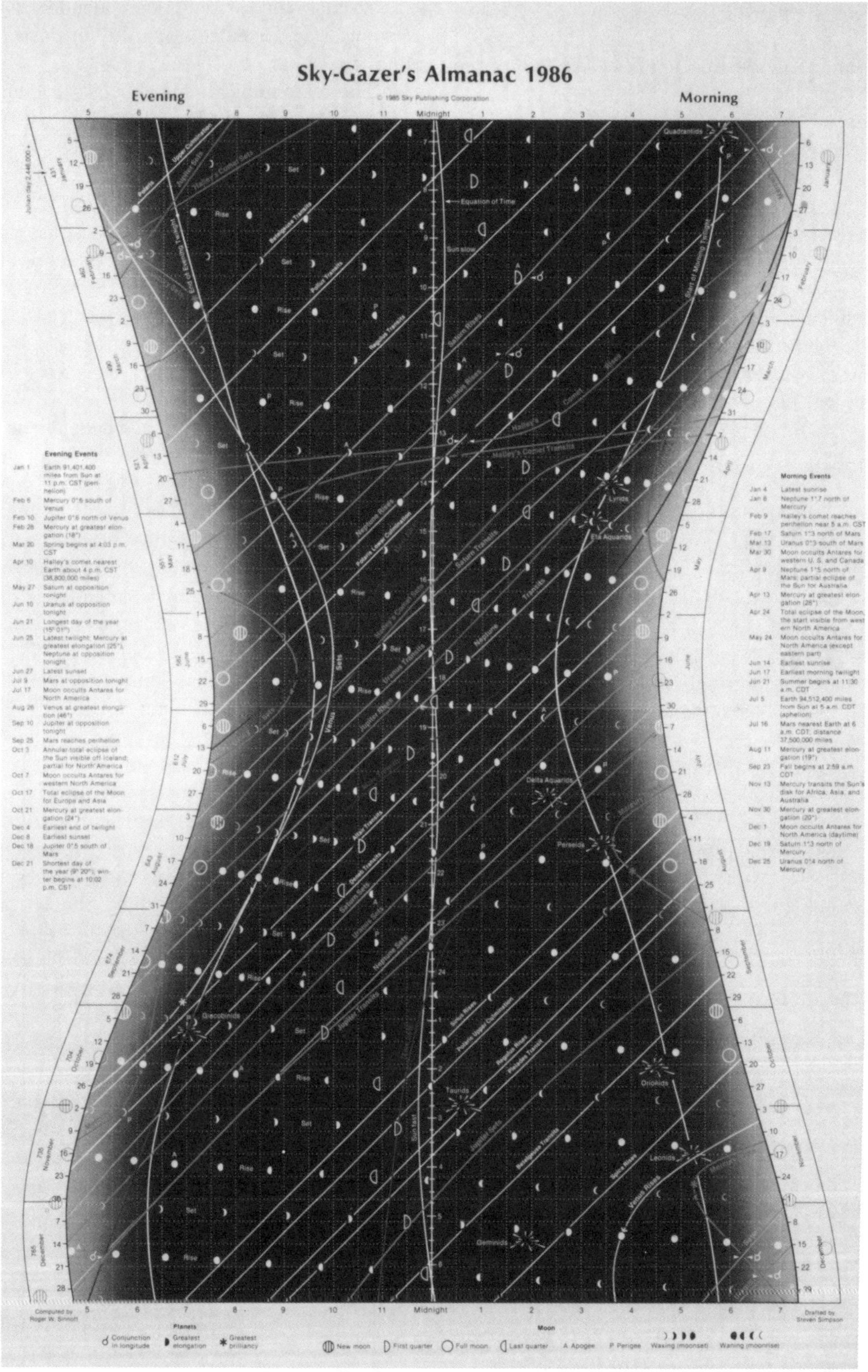

Figure 10.3 *This graphic almanac or "calendar" is from the January 1986 issue of* Sky & Telescope *magazine, and appears each year in the January issue. ©1985, Sky Publishing Corp.*

Appendix A

Definition of Terms

***:** symbol used in most computer languages to represent the multiplication operation.

A

ABS: function which returns the absolute value of its argument. `ABS`(5.6) returns 5.6 while `ABS`(-3.4) returns 3.4.

Altitude: the coordinate in the horizon coordinate system which measures how far up an object is from the observer's horizon.

Angle of Inclination: the angle formed by the ecliptic plane and the orbital plane of a solar system object.

Annular Eclipse: a solar eclipse in which the Moon's umbra does not reach the Earth. The effect is as if a smaller disk were placed in front of a larger disk.

Anomalistic Month: a month as measured from the time that the Moon is at perigee. This month is 27.5546 days in length.

Anomalistic Year: a year as measured from the time that the Earth is at perihelion until it is at perihelion again. This year is 365.2596 mean solar days in length.

Anomaly: the angle formed by an orbiting body, the center (or focus) of its orbit, and the orbital major axis. Also see mean anomaly and true anomaly.

Aphelion: the point in a planet's orbit at which it is furthest away from the Sun.

Apogee: the point in an orbit about the Earth at which an object is furthest away from the Earth.

Apparent Time: time as measured by the motion of the Sun. Apparent time is not uniform in length since it is based upon a solar day. Also see mean solar day.

Ascending Node: the point in an orbit at which the object rises above a reference plane. For the solar system, the ecliptic plane is the reference plane. Also see node and descending node.

Astronomical Almanac: a publication containing the location of various celestial objects, typically only solar system objects, and astronomical events of interest such as Sunrise and Sunset

Astronomical Telescope: a telescope in which the resulting image is usually inverted. Normally such a telescope is only used for astronomical observations where the inversion is unimportant.

Astronomical Unit: the length of the semi-major axis of the Earth's orbit. Typically astronomical units, abbreviated A.U., are used to measure distances within the solar system. One A. U. is about 9.29×10^7 miles.

Autumnal Equinox: the descending node for the Sun's orbit assuming the plane of the celestial equator as a reference plane. The Sun's right ascension at the autumnal equinox is 12^h. This occurs at about September 22. Also see equinox and vernal equinox.

Azimuth: the coordinate in the horizon coordinate system which describes how far around an object is from the North. This angle is measured eastward or clockwise from North.

B

Besselian Year: a year measured in the same way as the tropical year except that the starting reference point is when the right ascension of the mean Sun is 280°. This makes the Besselian year more closely correspond to the civil year. This year is 365.2422 mean solar days in length. Also see year, tropical year, and civil year.

Black Hole: a point in space where the gravitational field is so strong that even light can not escape. Thought to be formed by a collapsed massive star, black holes theoretically distort space and time within their realm of influence.

C

Celestial Equator: the great circle formed by the intersection of the plane of the Earth's equator and the celestial sphere. Also see celestial sphere.

Celestial Sphere: an imaginary sphere of infinite size on which the stars and planets can be considered as fixed. The Earth is normally the center of the celestial sphere. Also see celestial equator.

Celsius: the metric scale for measuring temperature in which water freezes at 0° C and boils at 100° C.

Centigrade: see Celsius.

Civil Time: time as measured by a mean solar day, from midnight, in which days are exactly 24 hours in length.

Civil Year: a year which is either 365 or 366 days in length depending upon whether or not it is a leap year. Some define a civil year to be 365.25 days in length. Also see Besselian year, leap year, and tropical year.

Clock Drive: a motor mounted on a telescope used to automatically compensate for the rotation of the Earth.

Conjunction: the moment when two objects have a common coordinate with the same value as viewed from a particular place such as the Earth. For example, the Moon is said to be in conjunction with the Sun at the time of the New Moon.

Constellation: any of the 88 groups of stars categorized by astronomers. Constellations usually receive their name after a mythological character (such as Orion), an inanimate object (such as the Big Dipper), or an animal (such as Canis Major, the Big Dog).

Culmination: the moment at which an object crosses the observer's meridian.

D

Dawn: the period of semi-darkness just before Sunrise. At this time the Sun is less than 18° below the horizon. Also see twilight.

Day: the time interval between two successive transits of an object across the observer's meridian. This period of time is around 24 hours depending upon how it is measured. Also see mean solar day, solar day, sidereal day, and mean sidereal day.

Decimal Format: a format for expressing time or angles as a single real number. For example, 12.567^h or 355.134°. Also see DMS and HMS format.

Declination: the coordinate in the equatorial coordinate system which corresponds to latitude. Declination is positive North of the celestial equator and negative South of the celestial equator. Also see right ascension and equatorial coordinate system.

Descending Node: the point in an orbit at which the object descends below the reference plane. For the solar system, the ecliptic plane is the reference plane. Also see node and ascending node.

DMS Format: a format in which an angle is expressed in an integer number of degrees, an integer number of minutes, and in seconds. For example, $36°40'39.2''$. Also see decimal format.

Draconic Month: a month as measured by using the Moon's ascending node as a point of reference. This month is 27.2122 days in length. Also see month.

DST: Daylight Saving Time.

E

Eccentricity: a measure of the "flatness" of an ellipse. Mathematically, the eccentricity is the ratio of the distance of the focus from the center to the length of the semi-major axis.

Eclipse: the total or partial obscuring of one body by another caused by the passage of one body into the shadow of or in front of another. In the case of the Sun and Moon, these are called solar and lunar eclipses. Also see solar eclipse, lunar eclipse, and annular eclipse.

Ecliptic: the Sun's apparent path in the sky; also the plane containing the Earth's orbit around the Sun.

Ecliptic Coordinate System: the coordinate system based upon the plane of the ecliptic. Ecliptic longitude is measured in degrees starting from the vernal equinox while ecliptic latitude is measured North or South of the ecliptic plane.

Ellipse: the collection of all points such that the sum of the distance of each point from the two foci is constant. This geometric figure is oval in shape when the two foci are distinct points but is a perfect circle if the two foci are the same point. Also see eccentricity.

Elongation: the angle formed by the Sun, the Earth, and the celestial body in question.

ET: Ephemeris Time.

Ephemeris: a table or publication containing the coordinate locations of solar system objects.

Ephemeris Time: a uniform time system defined by astronomers to account for irregularities in the Earth's rotation. Ephemeris Time is very close to Greenwich Mean Time.

Epicycle: a circle introduced in Ptolemy's model of the solar system to account for retrograde motion.

Epoch: an instant of time used as a standard reference point from which time is measured. Typically an epoch is stated for the measurement of equatorial coordinates.

Equation of Center: the numerical difference between the mean anomaly and the true anomaly. That is,

$$E_c = M - v.$$

Equation of Time: the difference between mean solar time and apparent solar time.

Equatorial Coordinate System: a coordinate system based upon the plane of the celestial equator. Also see right ascension and declination.

Equinox: either of two points at which the Sun crosses the celestial equator. These two points are the ascending and descending nodes for the Sun's orbit with respect to the celestial equator. During the equinoxes, the length of the day and the length of the night are very close to the same. Also see vernal and autumnal equinox.

F

f Ratio: the ratio of the focal length of an objective lens to the diameter of the objective lens. Also see focal length.

Fahrenheit: a system for measuring temperature in which water freezes at 32° F and boils at 212° F.

Fall: the three months of the year which begin on the date of the autumnal equinox.

Fall Equinox: see autumnal equinox.

Field of View: the portion of the sky visible through a telescope or binoculars, as measured in degrees. For example, a 4° field of view means that only 4° of the sky are visible through the viewing device.

First Point of Aries: see vernal equinox.

FIX: a function which returns the integer part of a number. For example, FIX(1.5) = 1 while FIX(−2.3) = −2. Also see INT.

Focal Length: the distance from the center of a lens to its focal point.

Focal Point: the point at which light rays passing through a lens converge.

Focal Ratio: see f ratio.

Focus: either of two points used in determining the points which lie on an ellipse. Also see ellipse.

FRAC: a function which returns the fractional part of a number without regard to the sign of the number. (Some computers do include the sign but as defined for this book, it is excluded.) For example, FRAC(1.2) = 0.2 while FRAC(−10.55) = 0.55.

G

Galactic Coordinate System: a coordinate system used for locating objects within the Milky Way Galaxy. This coordinate system is defined with respect to a plane which contains the Sun and the center of the Milky Way Galaxy.

Galaxy: a large-scale collection of stars and other objects which appear to form a cluster or group with a more or less visible structure. A typical galaxy contains an average of 100 billion stars and measures 1,500 to 300,000 light years across.

Gas Giants: the planets Jupiter, Saturn, Uranus, and Neptune.

Geocentric: centered with respect to the Earth. A geocentric universe assumes the Earth as the center of the universe.

Gibbous: the phases of the Moon in which the Moon is more than half illuminated but less than fully illuminated as seen from the Earth.

GMT: Greenwich Mean Time.

Great Circle: a circle of greatest possible diameter formed by intersecting a plane with a sphere.

Gregorian Calendar: the calendar system introduced by Pope Gregory in 1582 to reduce the errors in the Julian calendar system. This is the most widely used calendar today.

Greenwich Mean Time: the LMT for Greenwich, England. All other civil times are measured with respect to the GMT.

H

Heliocentric: centered with respect to the Sun. A heliocentric solar system assumes the Sun as the center of the solar system. Also see geocentric.

HMS Format: a format in which time is expressed as an integer number of hours, an integer number of minutes, and as a number of seconds. For example, $12^h13^m15.1^s$.

Horizon Coordinate System: a coordinate system based upon the horizon as seen by an observer. Also see azimuth and altitude.

Hour Angle: the difference between the local sidereal time and the right ascension of a body,

$$H = LST - \alpha.$$

Also see right ascension and local sidereal time.

I

Inclination: see angle of inclination.

Inertia: the tendency of a body to remain at rest or resist a change in velocity.

Inner Planet: a planet between the Earth and the Sun (Mercury and Venus).

INT: a function which returns the largest integer which is less than the given argument. For example, INT(1.5) = 1 while INT(−1.5) = −2. Also see FIX.

J

Julian Calendar: the calendar system introduced by Julius Caesar which divides the year into periods of 365 days except for every fourth year which contains 366 days. Also see the Gregorian calendar.

Julian Century: a period of 36525 days measured from January 0.0.

Julian Date: the number of days which have elapsed since noon GMT on January 1st, 4713 B.C.

Julian Day Number: see Julian Date.

K

Kelvin: a system for measuring temperature in which 0° K (−273.16° C) is the temperature at which there is no molecular motion.

L

Latitude: the distance, measured as an angle, that an object lies from the Earth's equator. Celestial latitude is measured from the ecliptic.

Leap Year: in the Julian calendar system, any year divisible by 4. In the Gregorian calendar system, any year divisible by 4 except for century years which must be divisible by 400.

Light Year: the distance that light travels in one year. One light year is about $5.87 \times 10^1 2$ miles.

LMT: Local Mean Time.

Local Mean Time: time measured with respect to the mean Sun at the observer's location.

Local Sidereal Time: the sidereal time for an observer which takes into account the observer's location on the Earth. LST is similar in concept to LMT except that LMT is based upon solar time while LST is based upon sidereal time. (Actually, sidereal time can only be local time, but LST is used here to distinguish from GST.)

Longitude: the distance, measured as an angle, that an object is from the Greenwich prime meridian.

Longitude of Ascending Node: the ecliptic longitude of the ascending node for an object within the solar system. Also see ascending node and ecliptic coordinate system.

Longitude of Perihelion: the ecliptic longitude of the point of perihelion for an object within the solar system. Also see perihelion and ecliptic coordinate system.

LST: Local Sidereal Time.

Lunar Eclipse: an eclipse of the Moon caused by the passage of the Moon into the Earth's shadow. Also see also eclipse and umbra.

M

Magnitude: a measurement, based upon a logarithmic scale, of the brightness of a celestial object.

Maria: dark areas on the surface of the Moon. Maria is derived from the Latin word for seas.

Mass: a quantitative measure of the amount of material in an object. Mass is closely related to weight and is often used (erroneously) as another word for weight. Mass does not depend upon gravity whereas the weight of an object varies depending upon the gravitational pull.

Mean Anomaly: the angle formed by an orbiting body, the center of an assumed circular orbit, and the orbital major axis. Also see true anomaly.

Mean Orbit: a fictitious orbit for an object that is assumed to be a perfect circle.

Mean Solar Day: a day as measured by the motion of a mean Sun.

Mean Sun: a fictitious Sun, moving in a mean orbit, defined in such a way that its motion is uniform. The mean Sun is defined so that a mean solar day can be measured.

Meridian: a great circle which passes through both poles of a sphere.

Modified Julian Date: the number of Julian days which have passed since November 17.0, 1858. Also see Julian date.

Month: the time interval required for the Moon to complete one orbit from one reference point to another. This is called a lunar month. The length of the month depends upon the reference point used to measure it. Also see anomalistic, draconic, nodal, sidereal, and synodic months.

N

Nadir: a point on the celestial sphere which is directly beneath the observer. Nadir is the opposite of zenith.

Nebula: a diffuse cloud of interstellar gas or dust that is visible as a luminous patch of dark and light areas.

Nodal Month: see draconic month.

Node: either of two points at which an object goes above or below some reference plane. Also see ascending and descending node.

Nutation: a small periodic wobbling in the Earth's axis of rotation.

O

Obliquity of the Ecliptic: the angle of inclination that the ecliptic plane makes with respect to the plane which contains the celestial equator; it is $23°26^m$.

Observer's Meridian: the meridian which passes through an observer's zenith and the North or South Pole.

Opposition: the moment when two celestial objects are directly opposite each other so that their longitudes are 180° apart. For example, the Moon is in opposition to the Sun when it is a Full Moon.

Outer Planet: the planets which are further away from the Sun than the Earth is. These are Mars, Jupiter, Saturn, Uranus, Neptune, and Pluto.

P

Parabola: a geometric figure formed by all points which are the same distance from a fixed line and a fixed point not on the line.

Parallax: the amount by which the apparent position of an object changes as the location of the observer changes.

Parsec: a unit of measurement based upon the distance from Earth at which the stellar parallax is one second of arc. One parsec is about 3.26 light years.

Penumbra: the lighter portion of a shadow in which not all of the light is blocked out. Also see umbra.

Perigee: the point in an orbit around the Earth at which an object is closest to the Earth. Also see apogee.

Perihelion: the point in an orbit around the Sun at which an object is closest to the Sun. Also see aphelion.

Planet: a celestial body, other than a comet or a meteor, in orbit around a star.

Power: a measure indicating how much a telescope magnifies an image.

Precession: a motion of the Earth's axis similar to the motion of the axis of a rapidly spinning top.

Prime Meridian: the meridian which passes through Greenwich, England.

Q

Quadrature: the moment when the angle formed by the Earth, a celestial object, and a reference point is $90°$. For example, the Moon is in quadrature during First Quarter Moon and Last Quarter Moon.

R

Refraction: a bending of light rays caused by a change in medium. For example, light rays are bent as they travel from space through the Earth's atmosphere.

Retrograde: an apparent backward motion of a celestial object as viewed from the Earth.

Resolution: a measure of how well a telescope is able to distinguish between two very closely spaced stars, or other fine detail in an image.

Right Ascension: the coordinate in the equatorial coordinate system which corresponds to longitude. Also see equatorial coordinate system, declination, and hour angle.

S

Saros Cycle: a period of approximately 18 years after which the cycle of lunar and solar eclipses tend to repeat.

Semi-major Axis: the length of the line segment which contains one focus of an ellipse, a point closest to the focus, and the center of the ellipse.

Sidereal Day: a day as measured by two successive transits of a star across an observer's meridian.

Sidereal Month: a month as measured by using a star as a reference point. A sidereal month is 27.3217 mean solar days in length.

Sidereal Time: time as measured from the instant when the vernal equinox crosses the observer's meridian.

Sidereal Year: a year measured as the time it takes for the Earth to return to the same position with respect to the stars. A sidereal year is 365.2564 mean solar days in length.

Solar Day: the time interval between two successive transits of the Sun across an observer's meridian. A solar day is not uniform in length. Also see mean solar day.

Solar Eclipse: an eclipse of the Sun caused by the passage of the Earth into the Moon's shadow. Also see annular eclipse.

Solar Time: time as measured by a Sundial. Solar time is uneven in length.

Solstice: either of the two points at which the Sun is furthest away from the plane containing the celestial equator. At the times of the solstices, the Sun has no apparent Northward or Southward motion. The longest day of the year occurs at the summer solstice while the shortest day of the year occurs at the winter solstice. Also see solstice and winter solstice.

Spring: the three months of the year which begin at the time of the vernal equinox.

Spring Equinox: see vernal equinox.

Star Atlas: a book containing a collection of star maps.

Star Catalogue: a book containing the coordinate locations of stars and other deep sky objects for some specific epoch. The catalogue may also contain other related data such as a star's magnitude.

Star Chart: a chart which shows the position of the stars with respect to each other.

Star Map: see star chart.

Summer: the three month period which occurs after the summer solstice.

Summer Solstice: the point at which the Sun is furthest North of the plane containing the celestial equator. This occurs at about June 22 and is the longest day of the year in the Northern Hemisphere.

Synodic Month: a month based on the mean interval between the same lunar phase: New Moon to New Moon, Full Moon to Full Moon, etc. The synodic month is 29.5306 mean solar days in length. Also see month.

T

Terrestrial Telescope: a telescope in which the resulting image is not inverted.

Transit: see culmination.

Tropical Year: a year as measured by using the vernal equinox as the point of reference. The tropical year is 365.2422 mean solar days in length. Also see civil year and Besselian year.

True Anomaly: the angle formed by an orbiting body, the Sun, and the orbital major axis. Also see mean anomaly.

Twilight: the period of semi-darkness just after Sunset. At this time the Sun is less than 18° below the horizon. Also see dawn.

U

Umbra: the darkest portion of a shadow in which all of the light is blocked out. Also see penumbra.

Universal Time: same as Greenwich Mean Time.

UT: Universal Time.

V

Vernal Equinox: the ascending node for the Sun's orbit assuming the plane of the celestial equator as a reference plane. The vernal equinox is also known as the First Point of Aries. The Sun's right ascension at the vernal equinox is 0^h. This occurs at about March 21. Also see equinox.

W

Waning: the phases of the Moon in which the amount of illumination is decreasing from Full Moon to New Moon.

Waxing: the phases of the Moon in which the amount of illumination is increasing from New Moon to Full Moon.

Winter: the three month period of the year which follows the winter solstice.

Winter Solstice: the point at which the Sun's position is furthest South of the plane containing the celestial equator. This occurs at about December 22 and is the shortest day of the year in the Northern Hemisphere. Also see solstice and summer solstice.

Y

Year: the time interval between two successive passages of the Earth or Sun past some reference point. The length of a year varies depending upon the reference point used. Also see anomalistic, Besselian, civil, leap, sidereal, and tropical years.

Z

Zenith: the point on the celestial sphere which is directly overhead an observer. Also see nadir.

Zodiac: a 16° wide band in the sky which contains the orbits of the visible planets.

Appendix B

Definition of Mathematical Symbols

h altitude. Range is $[-90°, 90°]$.

ω argument of perihelion (or perigee).

A azimuth. Range is $[0°, 360°]$.

δ declination. Range is $[-90°, 90°]$.

ϵ eccentricity.

β ecliptic latitude. Range is $[-90°, 90°]$.

λ ecliptic longitude. Range is $[0°, 360°]$.

ε ecliptic longitude at the standard epoch.

ϖ ecliptic longitude at perihelion (or perigee).

ε_g ecliptic longitude of the Sun at the standard epoch.

ϖ_g ecliptic longitude of the Sun at the moment of perigee.

Υ First Point of Aries.

l galactic latitude. Range is $[-90°, 90°]$.

b galactic longitude. Range is $[0°, 360°]$.

H hour angle. Range is $[0^h, 24^h]$.

ϕ latitude on the Earth. Range is [90°S, 90°N].

ψ longitude on the Earth. Range is [180°W, 180°E].

Ω longitude of the ascending node.

m magnitude.

M mean anomaly.

ε obliquity of the ecliptic.

ι orbital inclination.

α right ascension. Range is $[0^h, 24^h]$.

υ true anomaly.

THE GREEK ALPHABET

	Sign			Sign	
Name	Capital	Small	Name	Capital	Small
Alpha	A	α	Nu	N	ν
Beta	B	β	Xi	Ξ	ξ
Gamma	Γ	γ	Omicron	O	o
Delta	Δ	δ	Pi	Π	π or ϖ
Epsilon	E	ϵ or ε	Rho	P	ρ or ϱ
Zeta	Z	ζ	Sigma	Σ	σ or ς
Eta	H	η	Tau	T	τ
Theta	Θ	θ or ϑ	Upsilon	Υ	υ
Iota	I	ι	Phi	Φ	ϕ or φ
Kappa	K	κ	Chi	X	χ
Lambda	Λ	λ	Psi	Ψ	ψ
Mu	M	μ	Omega	Ω	ω

Appendix C

Suggested Readings

There are several references listed here that provide a more in-depth discussion of the topics briefly presented in this book. Of special interest are the books by Jean Meeus and Peter Duffett-Smith. These two books were the source for many of the algorithms presented in this text and contain more algorithms that might be of interest. All of the algorithms in the two books are suitable for implementation on a small microcomputer. Duffett-Smith is somewhat easier to read but the book by Meeus is an excellent value and contains a more rigorous mathematical development than Duffett-Smith.

For those interested in descriptive astronomy, the book *Asimov on Astronomy* by Isaac Asimov, although dated, is very entertaining. A must for any serious amateur are the books by J. B. Sidgwick. His works contain a wealth of information on telescopes and other astronomical equipment and virtually every aspect of astronomy that is likely to concern the amateur.

For the mathematically minded, W. M. Smart's textbook on spherical astronomy is suggested. *Fundamentals of Astrodynamics* is also recommended but be aware that it requires a substantial level of mathematical expertise.

The first two works cited below are available from the United States Government Printing Office on a yearly basis. Both are almanacs and are useful works to have when great precision is needed for some coordinates or when exact times are needed for some astronomical event.

Almanac for Computers 1989, Nautical Almanac Office, United States Naval Observatory. (published annually)

The Astronomical Almanac, Nautical Almanac Office, United States Naval Observatory, and Her Majesty's Nautical Almanac Office, Royal Greenwich Observatory. (published annually)

Astronomical Phenomena for the Year 1989, Nautical Almanac Office, United States Naval Observatory, and Her Majesty's Nautical Almanac Office, Royal Greenwich Observatory. (published annually)

Alexander, A. F., *The Planet Saturn.* Dover Publications, 1980.

Asimov, Isaac. *Asimov on Astronomy.* Anchor Press, 1975.

Barlow, B. *The Astronomical Telescope.* Wykeham Publishers, 1975.

Bate, Roger B., Donald D. Mueller, and Jerry E. White. *Fundamentals of Astrodynamics.* Dover Publications Inc., 1971.

Beatty, J. Kelly, Brian O'Leary, and Andrew Chaikin. *The New Solar System.* Sky Publishing, 1981.

Bell, L. *The Telescope.* Dover Publications Inc., 1981.

Belton, M. J. S. "Uranus and Neptune," *Astronomy*, February, 1980. p. 10.

Berry, *Build Your Own Telescope.* Charles Scribner's Sons, 1984.

Burnham, Robert Jr. *Burnham's Celestial Handbook.* Dover, 1978.

Chapman, C. R. *Planets of Rock and Ice.* Charles Scribner's Sons, 1982.

Cooper, Henry S. F. *Imaging Saturn: The Voyager Flights to Saturn.* Holt, Rinehart, and Winston, 1982.

Cunningham, Clifford J. *Introduction to Asteroids.* Wilmann-Bell, Inc., 1988.

Disney, Michael. *The Hidden Universe.* Macmillian, 1984.

Dobbins, Thomas A., Donald C. Parker, and Charles F. Capen. *Introduction to Observing and Photographing the Solar System.* Willmann-Bell, Inc., 1988.

Duffett-Smith, Peter. *Practical Astronomy With Your Calculator.* Cambridge University Press, 1981.

Gribbin, J. *Astronomy for the Amateur.* Macmillian, 1976.

Harrison, E. *Cosmology.* Cambridge University Press, 1981.

Harwit, M. *Discovery in Astronomy.* Harvester Press, 1981.

Hoyle, Fred. *The Nature of the Universe.* Harper and Row Publishers, 1960.

Illingsworth, *The Anchor Dictionary of Astronomy.* Anchor Press/Doubleday, 1980.

Jackson, Joseph H. *Pictorial Guide to the Planets.* Thomas Y. Crowell Company, 1973.

Jones, A. *Mathematical Astronomy with a Pocket Calculator.* Wiley, 1978.

Kaufman, William. *Planets and Moons.* W. H. Freeman and Co., San Francisco, 1979.

King, H. C. *The History of the Telescope.* Dover Publications, 1979.

Kingslake, R. *Lens Design Fundamentals.* Academic Press, 1978.

Lampkin, R. H. *Naked Eye Stars.* Gall and Inglis, 1972.

Markov, A. V. *The Moon: A Russian View.* University of Chicago Press, 1962.

Meeus, Jean. *Astronomical Formulae for Calculators.* Willmann-Bell, Inc., 1982.

Meeus, Jean. *Astronomical Tables of the Sun, Moon, and Planets.* Willmann-Bell, Inc., 1983.

Menzel, D. H. *Field Guide to the Stars and Planets.* Collins, 1977.

Middlehurst, B. M. *Chronological Catalog of Reported Lunar Events.* National Aeronautics and Space Administration, 1968.

Miles, H. G. *Artificial Satellite Observing.* Faber and Faber, 1974.

Moore, P. A. *Practical Amateur Astronomy.* Lutterworth, 1978.

Moore, Patrick. *History of Astronomy.* MacDonald and Company, London, 1983.

Moore, Patrick and Gary Hunt. *Atlas of the Solar System.* Rand McNally, 1983.

Muirden, James. *The Amateur Astronomer's Handbook.* Thomas Y. Crowell Company, 1968.

Muirden, James. *Astronomy with Binoculars.* Faber and Faber, 1976.

Muirden, James.*Beginner's Guide to Astronomical Telescope Making.* Pelham Books, 1975.

Peek, B. M. *The Planet Jupiter*, Faber and Faber, 1981.

Roth, G. D. *Astronomy: A Handbook.* Sky Publishing, 1975.

Roth, G. D. *Handbook for Planet Observers.* Faber and Faber, 1970.

Rutten, Harrie and Martin van Venrooij. *Telescope Optics Evaluation and Design.* Willmann-Bell, Inc., 1988.

Sidgwick, J. B. *Amateur Astronomer's Handbook.* Enslow Publishers, 1980.

Sidgwick, J. B. *Observational Astronomy for Amateurs.* Enslow Publishers, 1982.

Silk, J. *The Big Bang.* Freeman, 1980.

Smart, W. M. *Textbook on Spherical Astronomy.* Cambridge University Press, 1977.

Sullivan, Walter. *Black Holes: The Edge of Space, The End of Time.* Warner Books, 1979.

Sullivan, Walter. *We Are Not Alone: The Search for Intelligent Life on Other Worlds.* McGraw-Hill Book Company, 1964.

Texereau, J. *How to Make a Telescope.* Willmann-Bell, Inc., 1984.

Thompson, A. J. *Making Your Own Telescope.* Sky Publishing Corporation, 1947.

Tirion, Wil. *Sky Atlas 2000.0.* Sky Publishing, 1980.

Villard, R. "Do Asteroids Have Moons?," *Star and Sky*, September, 1980. pp. 17-24.

Webb, T. W. *Celestial Objects for Common Telescopes.* Dover Books, 1962.

Whipple, F. L. *Earth, Moon, and Planets.* Harvard University Press, 1963.

Whipple, F. L. *Orbiting the Sun.* Harvard University Press, 1981.

Wilson, Colin. *Starseekers.* Doubleday and Company, Inc., 1980.

Wood, J. A. *The Solar System.* Prentice-Hall, 1979.

Index

Other Items of Interest to the Celestial Calculator

Celestial Object Locater PACKage

Dr. Lawrence, the author of this book, has written a comprehensive software package COLPACK—Celestial Object Locater PACKage—designed to aid in locating celestial objects. It utilizes a simple menu-driven interface to make command selection quick and easy. A status line appears above each menu to show the current value of important variables such as latitude, longitude, and the date. This arrangement makes COLPACK behave much like an "astronomical spreadsheet." By changing one data item of interest (such as the Local Mean Time), dependent data items (such as Greenwich Mean Time and Local Sidereal Time) are automatically recalculated to account for the change. COLPACK also plots star charts and creates astronomical calendars.

COLPACK runs on IBM Personal Computers (PC, XT, or AT) or true compatibles with 256K RAM (512K recommended for large star data files); floppy or hard disk drive; DOS 2.0 or later; color or monochrome monitor with a Hercules, CGA or EGA graphics board. COLPACK will run on systems without a graphics board but it will not be possible to produce star charts.

Major features of COLPACK include the following:

- Time system conversions (LMT, GMT, LST, GST)
- Julian date conversions
- Date to day-of-week and day number
- Coordinate system conversions (ecliptic, equatorial, galactic, and horizon)
- Automatic calculation of coordinate corrections: precession, nutation, annual aberration, parallax (solar system objects only), and atmospheric refraction.
- Location of stars the Sun, Moon and the Planets
- Rising and setting times for the stars, Sun (including times of dawn and twilight), Moon and planets.
- Phase of the Moon and planets
- Distance to solar system objects
- Angular diameter of solar system objects
- Astronomical Calendar
- Star Charts
- Data File of Celestial Objects
- Orbital Elements (True & Mean Anomaly, Equation of Center, Argument of Perihelion, Orbital Inclination, Mean Motion per Day, etc.)
- Times of solstices and equinoxes
- Solar Elongation, Position Angle of Bright Limb, Equation of Time

See order form on next page.

Astronomical Formulae for Calculators

For the past 10 years Jean Meeus' *Astronomical Formulae for Calculators* has been the standard reference for basic astronomical formulae. While originally written for the programmable scientific calculator, many persons have found it a helpful aid to writing programs for microcomputers. Now in its fourth edition subjects covered are: Julian Day and Calendar Date; Date of Easter; Ephemeris Time and Universal Time; Geocentric Rectangular Coordinates of an Observer; Sidereal Time at Greenwich; Transformation of Coordinates; Angular Separation; Conjunction Between Two Planets; Bodies in Straight Line; Smallest Circle Containing Three Celestial Bodies; Position Angle of the Moon's Bright Limb; Precession; Nutation; Apparent Place of a Star; Reduction of Ecliptical Elements From One Equinox to Another; Solar Coordinates; Rectangular Coordinates of the Sun; Equinoxes and Solstices; Equation of Time; Equation of Kepler; Elements of the Planetary Orbits; Planets: Principal Perturbations; Elliptic Motion; Parabolic Motion; Planets in Perihelion and Aphelion; Passages Through the Nodes; Correction for Parallax; Position of the Moon; Illuminated Fraction of the Moon's Disk; Phases of the Moon; Eclipses; Illuminated Fraction of the Disk of a Planet; Central Meridian of Jupiter; Positions of the Satellites of Jupiter; Semidiameters of

Sun, Moon and Planets; Stellar Magnitudes; Binary Stars; Linear Regression and Correlation. *Mercury*, the Journal of the Astronomical Society of the Pacific said:

> In this little jewel of a book, Belgian amateur astronomer and astronomical calculator "par excellence" Jean Meeus brings together a host of astronomical formulae for observers, navigators, mathematical hobbyists, and anyone who enjoys putting a pocket calculator or home computer through its paces.

Fundamentals of Celestial Mechanics

This is an introductory text that should be accessible to a reader having a background in calculus and elementary differential equations. The original edition (published in 1962) has been radically revised, and emphasis is placed on computation. The numerical analysis needed for the computations is derived, and sample programs (run on a IBM-PC) are included. There are introductory chapters on the astronomical background and on vectorial mechanics. Sections dealing with the problem of two bodies include the use of universal variables, several methods (including that of Laguerre) for solving Kepler's equation, and three methods for solving the two point boundary value problem. The chapter on the determination of orbits includes two versions of Gauss' method, the application of least squares and an introduction to recursive methods. The chapter on numerical methods has been expanded, and includes three methods for the numerical integration of differential equations, one of which has full stepsize control. There are also chapters on perturbations, the three- and n-body problems, the motion of the Moon and the rotations of the Earth and Moon. The appendix includes numerical tables and derivations of properties of conic sections that are used in the text. The text includes several hundred problems, and suggested computer projects.

As minimum requirements, a user is assumed to have taken introductory courses in calculus, differential equations and mechanics. Other topics, such as numerical analysis, are developed in the text as needed. See order form at right.

Replacement Diskette for Introduction To BASIC Astronomy With A PC

Willmann-Bell, Inc. will replace defective diskettes for a period of 30 days after purchase provided that the diskette is returned postage paid (no COD's). Under no circumstances will Willmann-Bell accept return of a book for refund once the sealed envelope containing the program diskette is opened. Willmann-Bell does not accept return of books purchased from third parties—you must return the book and unopened diskette envelope to where you purchased them. Replacement diskettes are available directly from Willmann-Bell for $9.95 plus handling. See order form below.

Please send me the following:

☐ **Celestial Object Locater PACKage**	**$29.95**
☐ **Astronomical Formulae For Calculators**	**14.95**
☐ **Fundamentals of Celestial Mechanics**	**19.95**
☐ **Replacement Diskette for Introduction to BASIC Astronomy With A PC**	**9.95**
Handling[1]	**1.00**
TOTAL	________

I wish to pay with:
☐ **Check** ☐ **Money Order**
☐ **Visa** ☐ **MasterCard** ☐ **American Express**

Card No.________________________

Card expiration date________________________

Signature________________________

Name (Please Print)________________________

Street________________________

City, State, ZIP________________________

Willmann–Bell, Inc.
P.O. Box 35025
Richmond, Virginia 23235 U.S.A.
Voice (804) 320-7016 FAX (804) 272-5920

[1]Foreign orders: shipping charges are additional. Write for proforma invoice which details your exact costs for various shipping options.